THE JOURNAL
OF
BEAVER ISLAND
HISTORY

Volume One

*Essays on the history
of Beaver Island*

In this series:
THE JOURNAL OF BEAVER ISLAND HISTORY
Volume One
A Beaver Island Bicentennial Committee Publication
Reprinted by The Beaver Island Historical Society

THE JOURNAL OF BEAVER ISLAND HISTORY
Volume Two
Compiled, edited and distributed by the Beaver Island
Historical Society, St. James, Michigan 49782

Library of Congress catalog card number 81-642940
ISBN# 0-944216-01-3
(Series ISBN# 0-944216-00-5)

Preface

This book has been published by the Beaver Island Bicentennial Committee as a Bicentennial project. It was funded by a grant from the Michigan American Revolution Bicentennial Committee. It is being sold to raise funds for another Bicentennial project, the restoration of the old Post Office formerly adjoining the Mormon Print Shop, which is now used as the Beaver Island Historical Society's museum.

Those familiar with Beaver Island will no doubt recognize the authors in this volume from their past accomplishments. James Fitting has previously published a monograph on the island's archaeology. Robert Weeks has published numerous articles and a book on James Strang. His chapter in this book will appear, in a slightly different form, in *The Wisconsin Magazine of History*. Helen Collar has long been recognized as an authority on the island's history. Her essay on Mormon Land Policy in the *Michigan History* quarterly has become a classic, and her portraits of the early settlers adorn the museum's walls. David Gladish is the author of the local best-seller, *All In Favor Say Island*. Antje Price is the translator of the Protar diaries. Margaret Hanley was the island's reporter for five years. And Phil Gregg is the editor of the island's paper, the *Beaver Beacon*. The chapters by Mrs. Collar, Mrs. Hanley, and Mr. Gregg may soon appear in their own books.

The list of those who should be thanked for their constant support is too long to be given here: everyone in the community pitched in. But in particular we'd like to thank Phyllis Townsend, for having edited the manuscripts; Barbara Beckers, for having established the final format; Helen Collar, Frank Nackerman, John Gallagher, Archie LaFreniere, and Charlie Martin, for having read many or all of the papers; Carol LaFreniere and Karen Wojan, for their help preparing the manuscripts; A. J. Roy, the President of the Historical Society, for providing access to his materials; and Ed Wojan, the Executive Director of the Beaver Island Bicentennial Committee, for his help unearthing and recording research data, reading the essays — especially the last, which he veritably coauthored — making constant detailed suggestions, and in general, helping smooth out all the convolutions this project has entailed.

William Cashman
Project Director

Introduction

Beaver Island is like a poem whose simple language and apparently obvious meaning mask its abundant underlying themes.

The image the island presents to the majority of those who come upon it for the first time is one of vast, unspoiled natural beauty, a quiet and secluded haven detached from the mainland's frenzy. Hunters are delighted with the ample deer, rabbits, and game birds. Each fisherman has his special spot, either in Lake Michigan or in the numerous inland lakes, to which he returns year after year for trout, bass, or perch. Hikers are enchanted by the thousands of trails criss-crossing the fifty square miles of unfenced woods and fields. Botanists and biologists discover an unbelievable variety of flora and fauna here.

On the ferry back to "civilization" these people meet and exchange their visions. The picture that emerges from their shared experiences is one of an isolated paradise which has somehow escaped the inroads of progress. It's an attractive impression. But it's not the whole truth, as this book will show.

Those who hold this romantic picture might be surprised at how much historical research has been done on the island's past. Three reasons account for it. First, the research is a natural reaction to the implication that because the island is so peaceful now nothing must ever have happened here.

Second, Beaver Island is an ideal topic for research. Ninety-nine percent of what will ever be known about its past happened within the past century and a half. In contrast to those locales whose borders change from day to day, Beaver Island has remained constant for thousands of years. And Lake Michigan has acted as a buffer, reducing the effect of outside forces, so that the island has been exceptionally free to follow a course of its own design.

Third, and most important, is the history itself. The events had to have happened before the research could uncover them. They did happen, as this book will make clear. The clues are still with us, from the wrecked boats to be found along the beach, to the decaying farmsteads with their rusting old tools lying about, to the occasional ties and spikes to be found along the west bluff.

We hope this book will allow the reader to understand any of these clues he comes upon in his wanderings through this unspoiled paradise.

Downtown St. James with wooden sidewalks.

Table of Contents:

BEAVER ISLAND ARCHAEOLOGY

by James E. Fitting

With all of its exciting history, it is easy to forget that Beaver Island was occupied for thousands of years before the coming of the first Europeans and the beginning of recorded history. The tales of this earlier history are told in the land forms in the silent forests and in the broken pieces of prehistoric pottery and fragments of chipped stone that can be found by the careful observer in many parts of the island. These humble items provide the record of the island's unrecorded past, but for the archaeology of the island, like much of its history, fact has been clouded by legend as the fiction of what some would like to believe obscures what we really know.

There is no need to garnish the accounts of the island's past with speculations. They stand as an intriguing record themselves. The island first emerged from beneath the high water levels of the late glacial stage of Lake Michigan some time around 11,000 years ago. From then until the present, it was available for the use of man. The early Paleo-Indian hunters of barren ground caribou, and possibly walrus and whales, may have left their distinctive fluted spear points on the island but nothing so early has been found at this time.

The earliest traces of man are found along the Angeline Bluff on the western side of the island and may date to the time when this was an active beach of a higher lake stage. This bluff would have been an active beach between 2,000 and 2,500 BC and some of the "mounds" reported in this area appear to be fossil sand dunes which existed in back of this beach and have long since been stabilized by a cover of grass and trees.

Little is known about the people who lived along this beach. Their sites are small, temporary encampments which were probably utilized during the summer months. Their tool kits were meager although their very presence on the island demonstrates that they knew how to build and use canoes. They hunted on the island using throwing spears or darts, possibly with spear throwers or "atlatls." By inference, deer

1

were their most likely target although information from other parts of Michigan suggests that they hunted small game as well and they utilized what fish they could catch. They were living along the beach, now a bluff, almost 2,000 years before pottery vessels were first manufactured in this area. They cooked their food by roasting or "stone boiling," the dropping of heated rocks into skin lined pits. As a result, these sites are littered with fire-cracked rock and heat fractured pebbles which outnumber the chipped stone flakes.

The Indians who lived on the island before the introduction of pottery are known to archaeologists as "Archaic" peoples. Those who lived along the Angeline Bluff when it was an active beach belong to the "Late Archaic." The later period, when people made pottery, is referred to as the "Woodland Period." The Woodland Period in this area is divided into "Initial Woodland," lasting from perhaps 100 BC up to 700 AD, and the "Late Woodland" lasting from 700 AD until the time of European contact.

Initial Woodland remains have been found from the Traverse City area along the coast of Lake Michigan up to the Straits of Mackinac where sites of this period are quite common, particularly around St. Ignace. Initial Woodland pottery is thin and well made. Pottery vessels had pointed bases and were decorated with complex patterns made by stamping the wet clay with a comb-like tool, dentate tool, solid tool, or fingernail, before the vessels were hardened in open fires. The projectile points were spear or dart tips similar to those from the Late Archaic Period, suggesting that the bow and arrow had not yet come into general use. Small scraping tools, about the size and shape of the tip of a person's thumb, were common tools of this time.

Island and coastal sites of this time period are usually large summer fishing camps. They have been identified on most of the major islands and around the larger river mouths and bays in northern Michigan and Wisconsin. No remains from this time period have been reported from Beaver Island, which is surprising. Perhaps some of the mounds which have been reported on the island date to this time period though it is more likely that even these were built during the early part of the Late Woodland Period.

The mounds of Beaver Island need to be discussed in more detail. One of the early pioneers of Michigan archaeology, Henry Gillman, visited Beaver Island in 1871 and published an account of his visit in the *Smithsonian Institution Annual Report for 1883* (published in 1884).

"A remarkable series of mounds occurs at Beaver Harbor, on Beaver Island, in Lake Michigan. They are at present chiefly occupied by the town of St. James, which was built by the Mormons, under their leader, James Strang ("King Strang") about the year 1852-53. The mounds, which overlook the harbor, are extensive; and though, so far as I am aware, they have never been systematically investigated, they doubtless present a rich mine for research.

"A very limited and hurried examination which I made of them in 1871 sufficiently satisfied me as to their ancient origin. They appear to be of the same character as the mounds on the Detroit River and at the Foot of Lake Huron. They were probably largely used for purposes of sepulture, and until a comparatively recent period even the present race of Indians has continued to inter the dead, though not perhaps in the same repositories, at least in their immediate vicinity. From the success attending my brief labors, it would appear that the more valued relics of the "mound-builder" have been deposited in unusual abundance. Highly-wrought stone implements, many of them being of uncommonly skillful workmanship, are frequently encountered. These consist of axes, chisels, fleshing-tools, sinkers, spear-points, arrowheads etc., formed of a great variety of stone, such as diorite, sienite, greenstone, shale and cherts, many of them being finely polished.

"One of the handsomest stone-axes I have seen was taken out from this place. It was made from sienite, a favorite material for this implement, and the handicraft displayed in its construction is of a high order. Another axe, of diorite, is exquisitely polished.

"The implement (his Figure 10, a bipointed stone object two and a half inches long) found here and presumed to be a sinker, I have thought worthwhile forwarding a sketch of. The grooves shown at its middle and at each extremity, though shallow, are distinctly marked, and the entire implement is elegantly finished; apparently too much so by far for the purpose for which it is supposed to have been designed. It is made from a grayish shale, and is slightly polished.

3

"Another stone implement from those mounds is the large circular upper stone of the utensil conjectured to have been employed for grinding the grain used as food. This stone, which is also of sienite, is finely worked, being much smoother on one side than on the other. It is possible it may have been employed for another purpose than that suggested.

"Immense amounts of pottery (the well-known cord-pattern being frequent) were found here. In quality it compares favorably with that from the Detroit and St. Clair River mounds."

W. B. Hinsdale, in his *Archaeological Atlas of Michigan,* published by the University of Michigan in 1931, referred to Gillman's report. In Hinsdale's map, Gillman's four mounds have been reduced to two mounds and placed on Sucker Point. A burial site was located at the head of St. James Harbor with a village site just to the south of it. As an added note, Hinsdale listed additional prehistoric village sites on Indian Point in the northwest corner of the island and near Iron Ore Bay on the south end of the island. Surveys in the 1970s failed to locate either of these two villages.

As to Gillman's "mounds," they are not mounds at all but rather natural beach ridges or standlines such as are formed behind active beaches today. Gillman made this error in the Port Huron area as well as mistaking village debris for burial goods. It is interesting to note that Gillman reported no burials from his mounds although the rest of his article dealt with the characteristics of burials from real mounds in Michigan.

In 1973, I investigated the area between St. James Harbor, Sucker Point, and the old site of Pagetown rather thoroughly and found only one small mound, about 18 inches high and ten feet in diameter, in the woods to the north of St. James. This was not investigated since it had clearly been opened and rather badly disturbed in the past. The mounds on the Hinsdale map were located in the Whiskey Point area near where the St. James Township Hall (Coast Guard facility) stands today and, if they ever did exist, have long since been destroyed.

The richest archaeological site on Beaver Island is located in the place where Gillman found his artifacts. It underlies, and has largely been destroyed by, the modern community of St. James. It is still possible to pick up broken pieces of prehistoric pottery, flint chips, and an occasional arrow point, in the vacant lots and cleared areas of St. James.

4

The yard around the Mormon Print Shop may be the only undisturbed part of this large archaeological site and certainly should be preserved for scientific excavation some time in the future.

The Late Woodland pottery from this village is predominantly cordmarked with some slight decoration on the necks of the fragmentary vessels. Some pottery sherds have been found which have had crushed shell added to the wet clay. This is a characteristic of pottery from further to the south in Illinois and suggests trade with people to the south. Chippage is common but chipped stone tools are rare. Small, triangular projectile points are often found, indicating the use of the bow and arrow in hunting. While some hunting was certainly done, the scattered bone remains are primarily from fish. The pottery indicates that the most intensive period of occupation was from 700 AD to 1400 AD. There are similar, but smaller village sites of this same period located at several places along the east side of the island, including Sand Bay, Martin Point, and around Cable Bay.

I was drawn to Beaver Island by my interest in the archaeology of the 17th century, the time when the French were first entering the area and major changes were taking place in many facets of aboriginal culture. I had just finished my first two seasons at the Marquette Mission site and, although Marquette had not visited Beaver Island or included it on his map, there were other suggestions of French occupation.

The earliest of these references was in James Strang's *Ancient and Modern Michilimackinac,* first published in 1854 and based on articles which had appeared earlier in the *Northern Islander.* This account, with an introduction and comments by George May, was republished by W. Steward Woodfill in 1959.

"The French of Champlain's colony at Quebec were at Beaver Island before the Puritans reached Plymouth or the Dutch New York. Utensils left by them at different early periods are frequently found. Extensive fields which they cultivated are grown up to woods and some remain in grass.

"But there are strong indications of the presence of civilization at a still earlier period. The French settlement in Canada dates in 1608, but there are extensive fields on Beaver which have been thoroughly cleared and cultivated; and some very fine garden plots remain with the beds, paths and alleys as well favored as the day they

5

were made, and laid out on an extended scale; on which trees have been cut of two hundred and four years growth. Consequently, these places have been abandoned and grown up to timber, at least since 1650.

"There is room at least to believe that of the numerous European colines (sic) which were planted in America and lost without their fate ever being known, someone was carried captive to this recess of the continent, and allowed to remain in peace. The existence of such a fact is almost necessary to account for the rapid extension of Champlain's colony in this direction. For it is certain that within three or four years after Champlain commanded the colony at Quebec, it had extended to Beaver Island, and had a trading house at what is now St. James.

"In 1688 Baron LaHontan, Lord Lieut. of Placentia, passed this way on a voyage to and up the Saint Peter's River of Minnesota, near the head of which he found captives from the country around a salt lake beyond them, having beards, and the appearance of Europeans, which he took to be Spaniards; though they being slaves, and in the presence of their masters, called themselves Indians."

In his comments on the portion of Strang's account, May observed that, "there is no reliable evidence which would indicate any considerable French or other European settlement on Beaver Island in the 17th century." It was not mentioned in any of the Jesuit reports from St. Ignace and "Isle du Castor" did not appear on a French map until 1744. Still, this account continued to circulate and it has been included in books published in the last decade.

I have walked many of the clearings on Beaver Island and have yet to encounter a prepared ridged field system or aboriginal garden plot. Most of these meadows appear to be sand blows while others are clearly 19th century fields associated with Mormon farmsteads. There are certainly no signs of intensive 17th century occupation similar to that found in St. Ignace.

There is some suggestion of 17th century aboriginal occupation. An iron spear point of a 17th century style was found in the dunes near the mouth of the stream now called the Jordan River. A careful search was made of this area and no other cultural material was found. This is not a habitation site, rather an artifact probably lost while hunting.

6

Along the southern shore of St. James Harbor, a series of small rectangular fragments of chipped local flint have been found. These are early, pre-1670, gunflints of a type manufactured by the local Indians for use in their flintlocks. They are copies of gunflints made in Europe at the same time period only of local material. If the French were not on the island itself, their presence in the Great Lakes area, and its affect on local populations, is made manifest by these artifacts.

There seems to be little trace of 18th or early 19th century occupation on Beaver Island either. By the middle of the 19th century, however, there are archaeological manifestations of intensive settlement at both Cable's Bay and in St. James Harbor, as well as at small farmsteads all over the island. There are estimates of well over 2,600 people of the population in the 1850s, more than ten times the size of the modern population. While there are many scattered farmsteads, I would guess that Strang's population estimates, like his history, may have been both speculative and optimistic. The many scattered farmsteads, if properly surveyed and tested by scientific archaeological excavations, might be able to tell us more about the size, extent and economic activities of the Mormon population than all of the historical documents that have so far been uncovered.

The history of Beaver Island extends far back beyond its written history. The earth has started to yield its secrets of the past but has provided us with only a fleeting glimpse of what the past can teach us in the future.

THE KINGDOM OF ST. JAMES AND NINETEENTH CENTURY AMERICAN UTOPIANISM

by Robert P. Weeks

The Mormon kingdom established on Beaver Island in the 1850's by King Strang is usually depicted as a wildly eccentric scheme headed by a gifted con man who operated on the fringes of American society. There is much to support this view. Strang forged a letter from Joseph Smith naming himself as Smith's successor; he excommunicated Brigham Young, charging him with polygamy, then became a polygamist himself; he issued a strict dress code horsewhipping those who wore finery while he himself sported a crimson robe and a jewelled crown; he seized government lands and presented them to his followers as "inheritances" from God; and whenever he needed authority for a new policy, he conjured up a divine revelation.

But to see James Jesse Strang solely in these terms is to reduce him to a caricature. When we place Strang in the political, social, and religious context of mid-nineteenth century America, he is not on the fringes but very much in the American grain. For his kingdom illustrates the power of four distinct elements of our national experience: the experimental spirit of the Founding Fathers, the utopianism of various nineteenth century sects and socialist groups, the wild enthusiasm of American revivalism, and certain features of that most American of religions — Mormonism. The purpose of this essay is to place Strang and his kingdom in this context, thus adding both credence and meaning.

I.

In the early spring of 1835, James Madison, the fourth President of the U.S. and chief architect of the Constitution, agreed to be interviewed by the English writer Harriet Martineau. The interview took place in Madison's handsome house in Montpelier, Virginia. Madison, 86 years old and frail, was described by Martineau as "sitting in his chair with a pillow behind him; his little person wrapped in a black silk gown; a warm gray and white cap upon his head, gray worsted gloves" on his rheumatic hands.

9

Martineau writes that she sat "at the arm of his chair," for she was nearly stone deaf and used a hearing trumpet that consisted of an earpiece connected to a flexible tube that she handed to the person talking to her. She asked the elderly Madison about America's role in history. He placed the ivory cup of the hearing device to his lips and said of the country he had helped found that the United States was "useful in proving things before held impossible."

This theme of America as an experiment — and a successful one — or, as a kind of social laboratory for testing new communal arrangements, is a bright strand running through the fabric of our history. More than a half century before his interview with Martineau, Madison himself had defended the new Constitution against the attacks of his fellow Virginian Patrick Henry by declaring that "experiments must be made." In fact, of all the freedoms for which America stood, none was more significant, according to Historian Arthur Bestor, "than the freedom to experiment with new practices and new institutions."

With the Founding Fathers as example and inspiration, nineteenth century Americans extended the spirit of experimentation from political institutions to virtually every aspect of their lives: diet, religion, marriage and sexual customs, economic arrangements, and child rearing. They created hundreds of experimental communities, some with outlandish names: Ephrata, Nasoba, Zoar, Voree, Icaria, Nauvoo, Bohemia Manor, and The Woman in the Wilderness. Others had winsome names: Fruitlands, Equity, Harmony, Economy, Utopia, and Modern Times. All of them were expressions of the experimentation that was in the American air. It stirred first in New England, especially Massachusetts and New York, from which it flowed in a westerly direction through Pennsylvania and Ohio, eventually reaching every region of the country with the exception of the deep South. The two decades prior to the Civil War saw the most feverish activity. Shakers, Rappites, Owenites, Fourierists, Inspirationists, Perfectionists, and Associationists, and hosts of others launched their brave new worlds. In a famous letter to Thomas Carlyle in 1840, Emerson commented on this phenomenon: "Not a reading man but has a draft of a new community in his waistcoat pocket."

Just as the Founding Fathers were indispensable as models of the experimental approach, the great revivalists of the eighteenth and nineteenth centuries, for example Jonathan Edwards and Charles Grandison Finney, created an atmosphere of excitement, enthusiasm, and expectation that fostered the experimental mood. Nowhere did the flames of revivalism burn hotter than in northwestern New York, an area known as the "Burnt District" or "Burnt Over Region." Camp meetings and revival services drew hundreds or even thousands to a rural setting where marathon sermons were preached, sinners confessed their sins, and hundreds were converted in a frenzy of religious enthusiasm. Several months later, a new evangelist would appear, and the flames would leap again, Methodists, Baptists, and Presbyterians vying with each other in soul-saving showmanship.

Mormonism came out of the fires of the Burnt District. Its founder, Joseph Smith, grew up in the village of Palmyra, New York. His father, who switched from Universalism, to Methodism, then to Presbyterianism, and who had what he called visions, kept a small shop and did odd jobs on the side. Father and son spent much of their time hunting for buried treasure or lost articles with a forked stick, preferably of witch hazel and preferably in the summer; Joseph Sr. believed that the heat of the sun drew gold and silver coins to the earth's surface. They found little, although in digging a well, the son found a small, glasslike object that local superstition called a "peek stone," because it was supposed to enhance the vision and give its possessor supernatural power. In 1823, when he was seventeen years old, young Smith was startled by a visit to his bedroom in the middle of night. It was an angel named Moroni come to inform him that God had written a new Bible on golden plates and buried it in a nearby hill. Using his peek stone, young Smith located and dug up the plates and with the help of two peek stones provided by Moroni translated them as *The Book of Mormon,* and became the founder of the Church of Jesus Christ of Latter-day Saints.

In Scipio, another town in the Burnt District, James Jesse Strang, a farmboy of ten, was unaware that God, Jesus, and an angel named Moroni, had just visited Joseph Smith in nearby Palmyra; however, Strang, an alert, sensitive boy, was quite aware of the religious ferment in upstate New York. He has left a record of his feelings — his religious doubts, his ambitions, his philosophical musings — in a diary he kept as an adolescent and young adult.

Early in 1835, at approximately the same time that Harriet Martineau was visiting Madison, Strang, then twenty-two, wrote in his diary: "I have rejoiced in the sunshine and smiled in the shade of another year . . . It is gone . . . passed as others have passed their days who have died in obscurity." Then he added fervently, "Curse me eternally if that be my fate. I know it is in my power to make it otherwise." Men and women were being miraculously regenerated all around him, whole new societies were springing to life, the air was tonic with experimentation — and young Strang hankered to be part of it all. Three years earlier he had written: "I am 19 years old and am yet no more than a common farmer. 'Tis too bad. I ought to have been a member of the assembly or a Brigadier General before this time if I am ever to rival *Caesar or Napoleon* which I have *sworn* to." (The four underscored words were written in a cipher of Strang's invention not fully decoded until 1961.) Later that spring he confided to his diary in his private cipher: "I have spent the day in trying to contrive some plan of obtaining in marriage the heir to the English crown." (He refers to the future Queen Victoria, then twelve years old.)

When we consider these four critical elements of Strang's intellectual mileau — political experimentation, advocated by the likes of James Madison, utopian communities proliferating across the land, fiery revivalism an everyday occurrence, and dramatic divine intervention raising Joe Smith to the leadership of a major sect — the course of Strang's career becomes not only understandable but almost unexceptional.

II

After becoming a member of the New York bar and finding the law too tame, he wrote in his diary: "I have not seen enough of the world; played enough wild pranks, nor acted my part of its contention." Strang did what other Americans of his time and circumstances were doing: he headed for the frontier. During his first winter in the Territory, he attended a Mormon meeting in Burlington, Wisconsin, to hear an apostle of Joseph Smith's known as "The Wild Ram of the Mountains." When he saw how the Wild Ram could move his listeners to religious ecstasy, perhaps Strang was envious. We do know that he was sufficiently interested in Mormonism to journey two hundred miles to the south to Nauvoo, Illinois, where after talking to Joseph Smith he was baptized a Mormon.

As he baptized Strang, the Prophet said, "Thou shalt hold the Keys of the Melchizedek priesthood, shalt walk with Moses, Enoch, and Elijah, and shalt talk with God face to face." It was not quite like marrying the heir to the British throne, but it had possibilities. Especially in 1844; for Nauvoo seethed with political intrigue and violent anti-Mormon feeling. Clearly, the sect could not last much longer in Nauvoo. When Strang proposed to Joseph Smith and his brother Hyrum that he found a Mormon colony near Burlington, Wisconsin, they promptly made him an elder of the church and urged him to report on the possibilities to the north.

Strang's report from Burlingon came too late for Joseph Smith to do much about it. On June 27, a few days after receiving it, Joseph and Hyrum were killed by a mob in Carthage, Illinois. Smith had been such a colorful, powerful leader that to many his death meant the end of Mormonism. In the New York *Herald's* obituary, James Gordon Bennett wrote, "The death of the modern Mahomet will seal the fate of Mormonism. They cannot get another Joe Smith."

True, Smith was remarkable, but so was James Jesse Strang. In this awful crisis of the young church, only he responded with what amounted to joy as he announced that he had in hand a letter from Smith naming him as Smith's divinely chosen successor. And it was done with style: the beating wings of angels and "celestial musick" set the divine tone; Smith's signature and the dated Nauvoo cancellation clinched the authenticity.

The letter, folded and sealed within a sheet of paper postmarked "Nauvoo, June 19th," describes how Smith was "borne on wings of cherumbims" high above the Illinois countryside for a meeting with God who, predicting Smith's impending assassination, then pointed the divine finger at "james j. Strang" as the one who "shall plant a stake of Zion in Wisconsin . . . and there shall my people have peace and rest and shall not be moved for it shall be established on the prairie on white river in the lands of Racine and Walworth." Smith closes his letter to Strang, "if evil befall me thou shalt lead the flock to pleasant pastures."

It is not too much to claim that in the history of organized religion in the New World few letters have possessed the vast potential consequences of this one. The legitimacy of the leadership of a major American — and world — church hinges on the validity of this letter. If it is valid, then Brigham Young and his successors have defied the unmistakeably clear mandate of the founder of their religion; they are usurpers whose millions of followers have been duped.

13

Not surprisingly, James Jesse Strang exploited the letter to the fullest as documentary proof of the legitimacy of his claim. He carried it with him from town to town across mid-America, much as a medieval pardoner would carry his relics: a piece of the cross, a drop of the Lord's blood or one of the thorns from His crown. Prophet Strang exhibited the letter, he wrote in 1854, "in all principal cities and many of the towns and villages from the Mississippi to the Atlantic, and from the Potomac and Ohio to the Canada boundary." And he added that "in all that region, a hundred thousand witnesses are ready to bear testimony" as to its validity. Brigham Young was also keenly aware of the significance of that letter, and those pressing Young's claims to the succession did their utmost to prove it fake. And the battle still rages, the letter having been analyzed within the past dozen years by assorted experts. The controversy over the authenticity of the letter has focussed on the postmark, the handwriting, the paper, and the style.

When one looks at the letter today — it is preserved along with an immense body of other Strang material in the Coe Collection of the Yale University Library — there is at first glance little to arouse one's suspicions. It consists of two sheets of ordinary foolscap with writing on pages one, two, and three. The fourth page is the address-leaf, containing Strang's name and address, the postage rate, and the Nauvoo postmark handstamped in red ink. One might be somewhat suspicious in that the body of the letter is hand-printed whereas Smith's signature is script. But one might infer that Smith, a busy man, had the body of the letter written by a clerk, a common practice in the nineteenth century.

The present day verdict on the letter is that it is an ingenious amalgam. The postmark is genuine. The signature is not Smith's; moreover, it bears some significant resemblances to the handwriting of the body of the letter. One sheet of the paper — the part containing the address and postmark — was actually sent from Nauvoo, presumably by Smith. The other sheet, which does not match it, is clearly bogus. The style in which the letter is written bears only a superficial resemblance to Smith's. The conclusion: it is a fake letter constructed around a genuine sheet containing Strang's name and address.

To be a full-fledged prophet in the Joseph Smith tradition, Strang needed more than flights of angels and more than that letter: he needed a testament. This need was underscored by the fact that his chief rival to succeed Smith, Brigham Young, humbly acknowledged that he was no prophet. The most that the "Lion of the Lord" would say at that time was, "I am a good hand to keep the dogs and wolves out of the flock." As the battle of succession heated up, Strang set up a Mormon community named, in accordance with Smith's letter, Voree. He toured the favorite Mormon recruiting grounds — cities like Boston, Philadelphia, Washington, Baltimore, and New York — preaching and displaying Smith's letter. His first converts came from the collapsing Nauvoo; then the stream from the east began. Within a year of Smith's death, Strang presided over a semi-communistic community of a couple hundred faithful Saints in the rich farmlands of southeastern Wisconsin. But there was still no testament, no Strangite supplement to *The Book of Mormon.*

There were, however, as early as January of 1845, indications that if his followers were faithful, Strang would receive "the plates of the ancient records." In the fall of that year, he summoned four of the most reliable Saints — Aaron Smith, Jirah B. Wheelan, James M. Van Nostrand, and Edward Whitcomb — led them to an oak tree and told them to dig. When the blades of their shovels clanged against some metal plates, when these were found to be covered with the markings of an exotic language, when Strang received from an angel two stones with which he was able to translate this "lost Levantine" language, when he disclosed that it was holy writ from one of the lost tribes of Israel that had migrated to North America in 600 B.C., his followers were astonished. But the fact that the whole procedure duplicated Smith's discovery of *The Book of Mormon* twenty-two years earlier — the angel even providing Strang with Urim and Thummim, the two peek stones Smith had used — only seemed to validate the whole thing. At least for some.

Many of Strang's followers had experienced the fires of revivalism in New England, and most of them, we can assume, existed in what Christopher Lasch has described as a "chronic state of religious excitement" produced by the fierce competition among various sects. This state gave rise to two diametrically opposed states of mind. The unleashing of wild, religious enthusiasm caused some to draw back in doubt or even flat disbelief. (As a young man in the Burnt District, Strang seems to have had this reaction to religious excess, describing himself in his diary as "a cool philosopher" and even "an atheist.") Others were driven, according to Lasch, to the other extreme, to a search for ultimate religious truth, "a dogma to end dogmas." To these people, searching for finality and certitude, it would be profoundly satisfying to find a leader who with God's help vaulted over two millennia of religious dissension and disputation to get to the tap root of the whole thing: the tribes of Israel.

Many of the features of Mormonism derive in a clear and direct way from the superstitions of upper New York state, as Whitney Cross shows in *The Burned-Over District*. Treasure digging and stories of the lost tribes of Israel were both part of the local folklore long before the Smiths moved to Palmyra. But what gives Mormonism, in the Smith and Strangite forms, its deep appeal to Americans — whether of the 19th or the 20th centuries — is not its use of the folklore of upper New York but its skillful, even inspired, weaving of several strains of our national myth into a new, coherent American religion. The chief strains are: America is the Promised Land; the Indians are wicked apostates blocking fulfillment of our Manifest Destiny; those who are optimistic, hard working, and believe in the perfectability of man will enjoy success.

The earliest explorers of North America, the settlers, the poets and novelists who created our national literature, the clerics, the politicians, the statesmen, all have repeatedly invoked the myth of America as Eden or the Promised Land — a virgin land to be settled by God's chosen people. As widely held and as deeply lodged as this notion is in the American consciousness, it exists, nevertheless, as a metaphor, as an inspiriting and imaginative yoking of two otherwise dissimilar places: the Holy Land and the American wilderness. But Joseph Smith with a sublime combination of fundamentalist literal-mindedness and unbuttoned enthusiasm reduced the metaphor to fact: yes, this really *is* the Promised Land. His golden plates set the record straight: the Israelites were led here by God in 600 B.C. "They brought with them a record of all the Scriptures possessed by the Jews ... So great was their appreciation of the Almighty's present goodness to them that they kept a record of their own progress," according to the Mormons. But one group of Israelites split off, lost their faith, and became savages whose red skins were a sign of their apostasy. These Lamanites, as the Mormons call the Indians, destroyed the high civilization of their fellow Israelites, the Nephites, in 400 A.D. But like all American stories, this one has a happy ending. The Prophet will come, he will translate the Book of Mormon, with hard work and high expectations the Promised Land will bloom again, and even the Indians will get back their lost faith and become again, to quote Mormon scripture, "a white and delightsome people."

It is easy to see how this story would exert a strong appeal to the dispossessed, uneducated Americans who were looking for an authoritative, reassuring faith that provided relief from the fires of revivalism and a promise of a better future. Instead of meddling with the broad outlines of this story, Strang filled in the details. Through a series of divine revelations extending over several years, Strang created a blueprint for a new community. This was not a casual draft that would fit, as Emerson put it, "in a waistcoat pocket." More thorough and wide-ranging than the U.S. Constitution, *The Book of the Law of the Lord* ranges from Chapter I, The Decalogue, to Chapter XLVII, Payment of Debts. The most cosmic to the most trivial aspects of community life are codified: the religious hierarchy — priests, counselers, "embassadors," viceroys, apostles, etc. — is expansively set forth; civil government, getting down to such details as the length of women's dresses and the shapes of their shoes, is charted. If, as Lasch suggests, "The essence of Mormonism was the attempt to create a community of 'saints,' in which every 'secular' activity should be governed in accordance with a religious conception of the good society," then Strang's *Book of the Law of the Lord* is a central document for the understanding of Mormonism, even though it reads as if it had been co-authored by an Old Testament prophet and a New York lawyer.

Although Voree meant "Garden of Peace," it did not live up to its name. Some devout followers joined the communistic Order of Enoch and a secret order called the Illuminati whose members swore to uphold and obey Strang "as the Imperial primate and actual sovereign Lord and King on Earth." Others lost faith and either wandered off or made it their business to disabuse the faithful of the notion that Strang was either a prophet or a king. Local non-Mormons, or "Gentiles," as the Latter-day Saints dubbed everyone else, adopted the subversive practice of stopping the wagons of Strang's converts as they approached Voree from the east. With accounts of poverty, dissension, and chicanery within Voree, the Gentiles would try to convince the converts to turn back.

Voree could not long survive attack from apostates inside and Gentiles outside. Strang was learning the lesson that Alice Tyler said the leaders of many 19th century sectarian utopian communities had to learn if their communities were to survive more than a few months. The more peculiar the

tenets of a given sect, she wrote in *Freedom's Ferment*, the
more its followers needed to live apart in an isolated community in which to conduct "the intensive instruction, criticism,
and supervision that community living could make possible."
Brigham Young led his Saints to the deserts of Utah; King
Strang was visited by an angel who instructed him to abandon Voree and take his Saints to "a land amid wide waters."
The angel described it further as "covered with large timber,
with a deep bay on one side of it." Clearly, this was Beaver
Island and evidently the angel knew that the Indian claims
had already been invalidated and that shortly the government would open Beaver and the surrounding islands for settlement. Strang's kingdom was to be far more accessible than
Young's to the Mormon recruiting areas of the Midwest and
East; yet it was effectively isolated from the Gentile world by
the surrounding waters of Lake Michigan.

At nineteen Strang had measured himself against Caesar
and Napoleon and had confided to his diary in cipher, " . . . I
have learned all that I profess to know. That is that I am
ignorant and mankind are frail . . . I shall act upon it from
time to time for my own benefit." Certainly, to such a man a
kingdom on an isolated island was nearly as good as the
kingdom of heaven. And to many of his followers, Beaver Island was to become, indeed, a Promised Land. It was no accident that what the angel described as "the deep bay" was renamed by the Saints, "Paradise Bay."

III

Instead of transplanting Voree on Beaver Island, Strang
showed his considerable gifts as an adaptable, imaginative
leader in the changes he made. He dropped the two experiments with the communistic Order of Enoch and the secret
society of the Illuminati, shifting instead to "land reform" —
with a communitarian emphasis. In an illuminating article in
Michigan History, "Mormon Land Policy on Beaver Island,"
Helen Collar gives the essentials. Instead of relying on collec-
tivization as he did in part in Voree, Strang parceled out "inheritances" to the Saints who gathered there. Strang con-
ceived of the whole island as the Saints' inheritance from
God; he urged President Fillmore and Congress in April,
1850, "to pass a law giving the consent of the nation that the
saints may settle upon and forever occupy all the uninhabited
lands of the Islands of Lake Michigan and to cease to sell the
same to other persons." Not having heard from the politicians, Strang announced on July 8th on the occasion of his

coronation, that God had revealed to him that He had given the "Islands of the Great Lakes" to King Strang and his people.

Some of the parcels were paid for in the conventional way, but as Strang's political power increased with his election to the state legislature, he had his sheriff issue "certificates of sale," selling for nominal sums — or for nothing at all — land that belonged either to the government or to absentee Gentile owners. In addition to the capital city of St. James, five town sites were created. The King gave to each family an "inheritance" of 160 acres of farmland plus a small lot in St. James or the nearest of the six towns: Troy, Fontville, Siloum, Lowell, Gallilee, and Watamsa. His land policies were resourceful, quickly executed solutions to three problems faced by his kingdom: 1) the royal treasury, considerably dependent on tithing, lacked money for major land purchases; 2) the brief experiment in communism, the communal farm in Voree, had failed; 3) isolation, one of the key disadvantages of rural life, was underscored by the insularity of island life.

He solved the insolvency problem by bending or ignoring the law; for communism, he substituted cooperation combined with land reform; and he combatted isolation with a communitarian emphasis similar to that found in a host of 19th century utopian communities.

These policies seemed remarkably successful, and, indeed, they were — in the short run. But to the extent that they were illegal powerplays against Gentile landowners — a fact established by the Collar study — they contributed to the hostility that eventually destroyed the kingdom. The Mormons not only slyly outmaneuvered the Gentiles and expropriated land, but from the outset they openly re-named the lakes, rivers, hills, and other natural features on the island giving them Mormon or Biblical names as evidence that this land was their God-given inheritance. As offensive as these practices were to the Gentiles of northern Michigan, two other kinds of Saintly activity stirred up even more opposition to Strang's regime: the Saint's dealings with the Indians and their practice of polygamy.

In the 18th century, the pivot of the economy of northern Michigan was a tiny island at the tip of Michigan, Mackinac. It was the center of the fur trade for almost half the North American continent; then for a while it had served as a fueling station for wood-burning steamers. With the fur trade a trickle of what it had been and the island's timber all chopped

19

down, Mackinac in the mid nineteenth century existed largely on its past reputation as a trading center. Instead of fur, the region's economy was based on fish, timber, and whiskey; Beaver Island had a greater supply of the first two and a dependable demand for the third. As a teetotaler and an acute student of every facet of Beaver Island, including its economy, Strang was quick to see the significance of the fact that the Gentile trading post stood on Whiskey Point, a hook-like projection that formed the eastern rim of Paradise Bay. A major part of the Gentile trading activity consisted of bartering whiskey for fish caught by the Indians.

Strang traced the barrels of whiskey from the warehouses at Mackinac to the trading post at Whiskey Point, where they were used to produce what was called "Indian Whiskey." He wrote a forty-eight page pamphlet published with the Smithsonian Institution's Annual Report of 1854, which for half a century was the definitive work on the natural history, political life, and economics of the region. It was entitled *Ancient and Modern Michilimackinac, Including an Account of the Controversy between Mackinac and the Mormons,* and contained the following recipe for Indian whiskey: dump two gallons of common whiskey or unrectified spirits into thirty gallons of water; add red pepper to give it fire; and add tobacco to make it more intoxicating. The fish shipped from the rich grounds around Beaver Island had for thirty years been paid for largely with Indian whiskey, according to Strang's report.

Strang exposed this scandalous exploitation of the Indians in the Smithsonian pamphlet and in numerous articles in his newspaper, *The Northern Islander.* When the last Gentiles left the island in 1852, the whiskey trade had been cut off, and the Mormon fishermen, with better access to the fishing grounds than the fishermen based on Mackinac and the mainland, were taking over. Furthermore, with its superb harbor and extensive forests, Paradise Bay had become a popular fueling station for woodburning steamers. The Mormons were winning their economic war with the mainland; their political gains were impressive too. King Strang won re-election to the legislature in 1854 and began establishing colonies of Saints on the mainland near present-day Charlevoix.

Land acquisition, fishing rights, and trade with the Indians were basic causes of friction between Gentiles and Saints. All three were in various ways rooted in the theocratic groundplan of Strang's utopian community. The same is true

20

of the fourth and by far the most sensationalized cause of friction: polygamy.

For two thousand years, plans for utopian communities have included variations on monogamy. Plato, for example, argued that wives and families were potent causes of attachment to private property, so he proposed for one class in his *Republic* a community of wives. Early Christians, including John the Baptist and the disciples Peter and John, were familiar with a de-sexed version: a community of brothers and sisters. This became known as the institution of spiritual wives, a term adopted by the Mormons to describe polygamy, chosen perhaps for its genteel euphemism. In *Heavens on Earth,* Holloway traces polygamy in American utopian communities to these origins. But in their experimental zeal, American utopian communities were not limited to a single model; several, for example, adopted celibacy as a means of eliminating sexual possession and thereby solving the problem of exclusiveness. Whatever Strang's private motives, there can be no doubt that the American urge to experiment helped pave the way toward his adoption of polygamy; so did the more liberal attitude toward women that was, as Holloway puts it, "a logical consequence of the communistic attitude towards property." (Exactly how logical is suggested by Marx's dictum: "Marriage is incontestably a form of exclusive private property.")

Another precedent for polygamy lay much closer to Strang than any of these. Long before the establishment of the Mormon Church, various tribes of North American Indians practiced polygamy, including the Indians who lived in the Beavers before, during, and after Strang's reign there. Given Strang's deep interest in every facet of life on the islands — economic, political, geological, biological, and social — as well as his special interest in and respect for the Indians, or Lamanites, it is not unreasonable to assume that he became aware of the polygamy they practiced. (In fact, so great was Strang's interest in the Indians that when he first referred to the plan to move to Beaver Island from Wisconsin, he gave the project the heading "Indian Mission.")

Moreover, the descriptions we have of polygamy as it was practiced on Garden Island during the nineteenth century indicate that in certain respects it was quite similar to the arrangements Strang promulgated. For example, the Indians required that the tribe grant its approval after a male demonstrated his ability to provide for two wives. Strang wrote the same principle into *The Book of the Law of the Lord,* Chapter LXVI, section 5: "Thou shalt not take unto thee a multitude of wives disproportionate to thy inheritance, and thy substance . . . " We lack incontrovertible proof that he was influenced by the Indians; nonetheless, the evidence is strong that their example contributed to the various forces pushing Strang toward the adoption of polygamy.

Despite these precedents, Strang's path to polygamy was full of theological twists and extensive detours designed to put as much distance as possible between him and the Brighamites. In an editorial entitled "Polygamy Not Possible in a Free Government," he levelled against polygamy the two ultimate charges: it was heretical and undemocratic. But with the Brighamites settled in Utah and Strang's kingdom solidly established on Beaver Island, the King evidently felt that the propitious moment had arrived for him to give in to the forces pulling him toward polygamy. It is futile to speculate on the degree to which his own concupiscence was part of these forces; what is far more important to our purposes is to point out that polygamy was quite consistent with the communitarian and communistic values of his kingdom. Like the celibacy of the Shakers and Rappites, the free love of the Perfectionists, and the other sexual arrangements being experimented with in a variety of utopian communities in the 1840's and '50's, Mormon polygamy on Beaver Island was a local response to a general set of conditions. In the sexual conventions of his community — as in its religious, economic, and political arrangements — Strang acted not arbitrarily but in response to the general air of experimentation, especially as those were influenced by revivalism and Mormonism.

If Strang had not been shot down by a pair of disgruntled Saints on June 16, 1856, it is entirely possible that the kingdom could have survived for many more years. But once the leader had fallen, the hostility that his political, economic, and religious practices had generated among the Gentiles capsized the kingdom. If Strang had relied on a dozen able disciples, as Smith had, or if he had even designated a successor in the several days between the assassination attempt and his death, the kingdom might have survived its king. Since he did neither, the collapse occurred in short order. Lack of leadership on the island was increasingly clear to Gentile raiding parties from Mackinac, and they became bolder. The climax came July 5th when a mob, largely from Mackinac, arrived to drive the Mormons from Beaver Island. Bands of half-drunk

armed men roamed the island herding Mormon farmers and
their families at gun point to the dock in Paradise Bay and
aboard waiting steamers. One ship took 490 to Chicago;
others were dropped off at other Great Lakes ports — Detroit,
Cleveland, Buffalo. The dispirited Saints were so over-
whelmed that none resisted to any extent. Within the span of
a day or so, 2,600 men, women, and children were ruthlessly
uprooted and cast out. One reputable Michigan historian,
Byron M. Cutcheon, has called the fifth of July, 1856, "The
most disgraceful day in Michigan history."

IV

If this effort to place James J. Strang in the political, social and religious context of mid-nineteenth century America
has succeeded, then instead of seeming to be a freak — a sultan in a backwoods harem, an Old Testament Prophet
washed up on an island in Lake Michigan — he will be seen,
to a large extent, as a product of his extraordinary times.

It does not follow, however, that his contemporaries perceived him in this light. One of the most durable paradoxes of
history — as well as one of its strongest justifications — is
that events, their causes and effects, are often perceived most
clearly not by those who experience them at first hand but by
those who study them years later. Although it is somewhat
difficult for us, one hundred and twenty years after his death,
to assess the significance of Strang, for many of his contemporaries it was almost impossible. This can be illustrated by
an episode that was one of the triumphs of Strang's career. It
occurred because the President of the United States was unable to perceive Strang for what he was, and because a United
States District Attorney thought Strang was an evil sultan
out of the Arabian Nights.

During the spring of 1851, President Millard Fillmore
was visiting his brother in Detroit. He read the Detroit *Free
Press* and the Detroit *Advertiser and Tribune,* two papers
that found King Strang irresistible. Evidently, the Whig President put enough credence in the lurid accounts of Mormon
sexual debauchery, thievery, counterfeiting, and political
abuses to think that they could be used to send a popular
Democratic state legislator to prison.

Accordingly he authorized placing the U.S.S. *Michigan* at
the disposal of the U.S. District Attorney, George C. Bates.
Armed with warrants charging Strang and thirty-eight of his
top officials with treason, counterfeiting, trespass, theft, and

several other crimes, the district attorney, accompanied by a marshal and a judge sailed into Paradise Bay on the *Michigan* one night in May. The district attorney stealthily led a body of sailors from the *Michigan,* each armed with a Navy revolver and cutlass, toward Strang's house, their way lit by a covered ship's lantern. After stationing a boatswain at each end of the house, Bates, according to his sworn testimony, crept up the stairs until he found himself "in a long, low room, where wide berths, heavily draped with stunning calico, shielded beds like the berths and staterooms of steamers, which proved to be occupied by Mormon women four in a bed."

This account of the raid on the seraglio by the intrepid D.A. is mostly romantic nonsense. At no time did the king's palace have a harem or anything resembling what Bates described. We have a firsthand report by a Gentile observer of the living arrangements in Strang's house, and it bears no resemblance to what Bates says he saw. At the time of the raid, Strang's wife, Mary, was in Wisconsin and Strang was living in the house with twenty-year old Elvira, the only other wife he had at the time. A jury of ten Whigs and two Democrats acquitted Strang and his thirty-one co-defendants of all the charges brought against them. It was a major triumph for Strang.

This is a suitable episode with which to conclude this brief discussion of Strang and his kingdom because it provides us — admittedly in extreme form — with the circumstances under which a highly biased report of one aspect of Strang's community was made. It also provides us with the results of that report: a wildly imaginative picture of a Turkish harem transplanted to the north woods, illuminated by a covered ship's lantern and seen through the eyes of a politically inspired D.A.

This essay has attempted to see Strang and his kingdom not as a lurid sideshow — even though there are undeniable circus elements in his career — but as an integral part of our national experience. Seen in this way, the Kingdom of Saint James becomes a response to the experimental spirit of the Founding Fathers, an example of American utopianism, an episode in our bout with the fevers of revivalism, and a chapter in the development of a major American religion. It becomes, in other words, more believable and more meaningful.

BIBLIOGRAPHY

Bestor, Arthur. *Backwoods Utopias: The Sectarian Origins and the Owenite Phase of Communitarian Socialism in America: 1663-1828*. Philadelphia, 1970.

Collar, Helen. "Mormon Land Policy on Beaver Island," *Michigan History*, Summer, 1972.

Holloway, Mark. *Heavens on Earth: Utopian Communities in America, 1680-1880*. London, 1951.

Kanter, Rosabeth Moss. *Commitment and Community: Communes and Utopias in Sociological Perspective*. Cambridge, 1972.

Lasch, Christopher. *The World of Nations: Reflections on American History, Politics, and Culture*. New York, 1973.

Martineau, Harriet. *Retrospect of Western Travel*. London, 1838.

Noyes, John H. *History of American Socialisms* (formerly *Strange Cults and Utopias of 19th-century America*). New York, 1961.

Quaife, Milo M. *The Kingdom of Saint James: A Narrative of the Mormons*. New Haven, 1930.

Strang, James Jesse. *Ancient and Modern Michilimackinac*. St. James, 1854.

Strang, Mark. *The Diary of James J. Strang, Deciphered, Transcribed, Introduced and Annotated*. East Lansing, 1961.

Tyler, Alice Felt. *Freedom's Ferment: Phases of American Social History to 1860*. Minneapolis, 1944.

Weeks, Robert P. *King Strang: A Biography of James Jesse Strang*. Ann Arbor, 1971.

26

IRISH MIGRATION
TO BEAVER ISLAND

by Helen Collar

Irishmen lived on Beaver Island before the Mormons established the ill-fated Kingdom of James J. Strang. When Strang landed in 1847 to make his preliminary survey, there were already two settlements, one at the harbor and another on Cable's Bay. Living at the harbor was Patrick Kilty. He had first seen the island when, in 1845, he had been a member of the government surveying party which prepared Beaver Island land for sale to the public. Kilty had been living on Mackinac, but after the survey he returned to Beaver Island. His aunt and uncle, Patrick and Ann Luney of Toronto, Canada, came to visit their nephew in his new home and were so pleased with what they found that they too decided to settle on the Michigan island. All three had been born in Ireland.

The land which Patrick Kilty had helped to survey was put up for sale on July 17, 1848. The records of those earliest sales give us our first certain information about Beaver Island residents. One of the first to buy was Patrick Luney. He paid $54.62 for 43.7 acres of land on the point which still bears his name. Kilty bought no land at that time, but another Irishman, whose name is well known in island history, made his first purchase that same month. During the summer of 1848, Thomas O'Grady Bennett, who later died by Mormon gunfire, bought 140 acres fronting on Little Sand Bay.

The next definite records we have of Irish settlement are contained in the 1850 United States Census. When the census taker reached Beaver he found that the island had a population of about 483 people, 74% of whom were Mormons. The Mormons were living in the town of St. James at the harbor and on farm land in the interior. The only non-Mormons in the northern part of the island lived in town. They were Peter McKinley and his family, owner of the trading post on Whiskey Point, and two resident managers of the Northwest Trading Company of Rochester, New York, Erri James Moore and Randolph Dinsmore. This company's store and warehouse were on the west side of the harbor, not far from the present Parish Hall.

To complete his enumeration the census taker went to Cable's Bay. At the northern end of the bay were the store and dock of James Cable, and here he found a village of 33 households numbering 128 people. The Cables, James and his uncle Alva, had been the first white men to establish permanent residence on Beaver Island. Alva had originally built the property on Whiskey Point in 1838, held, in 1850 by Peter McKinley. After the arrival of a large number of Mormons, Alva joined his nephew at Cable's Bay where James had earlier set up an independent business. Among other residents of the southern village, the census taker found the Luneys, Patrick Kilty, the Bennett brothers, Tom and Sam, as well as 15 other people who had been born in Ireland.

Although Patrick Luney and Tom Bennett, like Alva Cable, had bought land at the harbor in 1848, we find that by 1850 they had deemed it prudent to leave their land and the homes they had built in order to move to the more friendly atmosphere of Cable's Bay. Beginning in 1848 the migration of Strang's Mormons had increased rapidly, and by 1850, as we have seen, they far outnumbered the original settlers. From the first, there had been friction between the Mormons and those whom they called Gentiles, and, as Strang's Saints multiplied their numbers, their power increased. By 1850 they were in control of the town of St. James, and all non-Mormons, except Peter McKinley and the two resident agents of the Northwest Company, had left the harbor area. These three men had heavy investments in their trading posts, and were evidently reluctant to leave their property to the mercy of the Mormons, particularly as Strang had had a revelation that non-Mormon property should be "consecrated" as "spoil unto God." Two years after the 1850 census was taken, Mormon power had become strong enough so that King Strang was able to pressure the non-believers, both at Cable's Bay and at the harbor, into abandoning their homes. Of the 146 people thus expelled, 102 were adults, and of these 23 had been born in Ireland. In the fall of 1852 all Gentiles had left Beaver, and the Mormons were in sole possession of their Island Kingdom.

After Strang's assassination and the driving away of his followers in 1856, many of the exiles of 1852 returned, among them Peter McKinley to the harbor and James Cable to his property on Cable's Bay. Some of the Irish exiles also returned, Mr. and Mrs. Luney, Patrick Kilty, Patrick Carmody, and Patrick Sullivan. These natives of Ireland all became prominent members of the Irish community that was to de-

28

velop on Beaver Island. Luney was appointed second keeper of the lighthouse at the Head; Kilty set up his fishing business near the point that still bears his name; Carmody, also a fisherman, married Big Mary, sister of Vesty McDonough, one of the first to come to Beaver after the Exodus; and Patrick Sullivan lives on in the records as part-donor of the land on which Beaver Island's first parish church was built, the church of St. Ignatius. St. Ignatius, later abandoned, stood in the clearing just north of the house we now call "Shoemaker's." The returning exiles were joined by others and a fishing community quickly developed, predominantly Irish in origin.

Before we go further, an explanation must be made of the distinctive system of nomenclature used on Beaver Island. I have been told that, at the same time, there were once living on Beaver nine Anthony O'Donnells and eight Hugh Boyles and as neither the Boyles nor the O'Donnells were as numerous as the Gallaghers, it is easy to see what confusion could result. It is obvious that if conversation was to be intelligible, some way of modifying the names of individuals had to be developed, and it was only natural that a system of nicknames arose that had its roots far back in the Islanders' Irish heritage.

Sir Henry Piers wrote about the Irish in 1683:

They take much liberty, and seem to do it with much delight, in the giving of nicknames: if a man have an imperfection or evil habit he is sure to hear of it in a nickname. Thus, if he be blind, lame, squint-eyed, left-handed, to be sure he will have one of these added to his name; also from the color of his hair, as black, red, yellow, brown, etc. and from his age, as young or old, so that no man whatever can escape a nickname who lives among them or converses with them.

This quotation is an accurate description of the system of nicknames used on Beaver Island. There was a Paddy Bacah. Bacah in Gaelic means crippled, and on inquiry I found Paddy had indeed been lame. Darky Mike, who gave his name to a road, was so called because of his swarthy skin, and Black John Bonner had black hair and beard. It is interesting that at times the Beaver Islanders kept the Irish word, and at others they translated it into English. For example; one red head was called Paddy Ruah, but another was known as Red Hughie; Young James was so called to distinguish him from his father, but there was also Katcheline Og, the Og meaning

29

young and Katcheline a diminutive of Katherine. There was another Katcheline on the Island, Katcheline Mor, Mor means either old or large, and in some others this nickname was Anglicized to Big as in the family of brothers, Big Dominick, Big Owen, and Big Neal.

Nicknames became patronymics; there was Big Owen's daughter, Mel-Big-Owen, pronounced as one word, Frank Danny Barney was Frank, the son of Danny Barney, the son of Barney O'Donnell for whom Barney's Lake was named. Paddy Mary Ellen and Andy Mary Ellen were sons of one of the best known mid-wives on the Island. In discussing patronymics it is interesting to note that Kilty means left-handed so the Patrick of the surveying party must have had a left-handed ancestor whose nickname became a surname. In the following discussion of the Irish who lived on Beaver Island I will use, for the sake of clarity, and without quotation marks, the nicknames that were more a part of a Beaver Islander's personality than the name he had been given at the baptismal font.

The first Irishman to reach Beaver Island after the shooting of King Strang was Black John Bonner. In the summer of 1856 he had established a fishing camp on Gull Island. One morning an Indian landed his canoe at the camp and, jumping ashore, shouted, "Big man shot! Big man shot!" Black John immediately took off in his boat for the 'big man's' island kingdom to find out what had happened. Although Bonner, at the time of the attack on Strang, was temporarily living on Gull Island, he was part of an Irish colony of fishermen which had developed on Mackinac in the waning days of the fur trade. This colony was made up of people, most of whom were fishermen, who had come to America from Ireland's County Mayo. As the best fishing in the northern lakes was to be found around the Beavers, they had been setting their nets in that area.

Feeling had been running high between those fishermen and the Mormons. Two Mackinac brothers from County Mayo, James and John Martin, had been set upon when, because of bad weather, they had beached their boat on the north shore of Beaver Island. Soon surrounded by a party of Mormons, they had been stripped of their gear, their clothes down to their underwear, and their oars, and then had been set adrift in an off-shore wind. The weather was freezing. At the last minute a Mormon, Harrison Miller, had thrown one oar back into the boat. Aided by this oar, and a providential change in the direction of the wind, the brothers managed to reach Garden Island. There they were helped by friendly Indians, and James and John got back to their homes on Mackinac.

When the news of the shooting of the King reached Mackinac Island and the nearby mainland, the many enemies of the Mormons made plans to eliminate the hated settlement, now leaderless. A call for rendezvous went out from St. Helena's Island where the Newton brothers had a trading post. Under the leadership of Archie Newton, a miscellaneous group gathered, made up of traders, fishermen, and off-duty soldiers from Fort Mackinac. Chartering the schooner "C.L. Abel," owned and captained by John Waggley of Mackinac, they set sail with, according to one account, "80 armed men." This was the group who drove the Mormons from Beaver Island. How many of these men were Irishmen who later made their homes on Beaver Island we do not know.

Provoked by depredations in the fisheries, and by the forced exile of the Gentiles in 1852, the landing party deported the Mormons with all the accompanying cruelty which is indigenous to mob actions. Group memory often erases actions such as these, and most Island families have no tradition of their grandfathers having been a part of the mob. It is known that Patrick Carmody was in the group, and Joe Palmer, a Scotchman who had lived on the island in pre-Mormon times, and whom we know returned with his family immediately after the Exodus, claimed, in later years, to have "helped drive off the Mormons."

Also in the group was Lewis Gebeau, the Indian half-brother of the later keeper of the Harbor Light, Elizabeth Whitney Williams. We read with wry amusement in her book "Child of the Sea" that her brother was "one of the men chosen to help preserve order in the sending away of the Mormons after the King was shot. He went to the Island to help get the people away on the steamboats that were sent to carry them from the Island." The hapless Mormons would undoubtedly have been glad to dispense with brother Lewis's help. Group memory not only conveniently forgets, it also distorts.

Except for these three men, plus two Newton brothers and Captain Waggley, the make-up of the armed and self-appointed avenging party is unknown. What we do know is that shortly after the Mormon Exodus several of the Mackinac Irish moved to Beaver Island. Black John Bonner, as we have seen, was the first to land. He soon brought over from

31

Gull Island his bride of a few months and bought the land on which the Parish Hall now stands. The next year, 1857, Bonner purchased the Michigan Centennial farm still owned by his descendants. It was there that he established his permanent home, the house built from logs that had been prepared by the Mormons, and found by John at Gravelly Run on the Fox Lake Road. Johnny the Rat O'Donnell (he earned his nickname by his exceptional agility in climbing the ratlines of sailing vessels) must have either come with Bonner or followed soon after. Johnny was Bonner's fishing partner.

Among the earliest of the other Mackinac Islanders to move were the Martins. This family included not only James and John who had been set adrift by the Mormons, but also their three brothers, Michael, Daniel, and Edward. The Martin boys had the distinction of not only being refugees from the Irish Famine, but also from the law. They had "broken the pound" at Westport; that is, after breaking into the pound at night, they had returned to their owners the cattle seized by authorities in lieu of debts owed the government. In poverty stricken Ireland this was a common, and honored, offense applauded by all who suffered under the hated English rule. The brothers were condemned and sentenced to deportation to Van Diemann's Land, but before the sentence could be carried out they were rescued by friends who "broke the jail," and got the brothers onto a ship bound for America. Eventually the Martins reached Mackinac Island, and there James and Edward married, Edward to Grace Malloy, and James to Katherine McCarty, sister of Patrick Kilty's wife Ann. The Martin men came over to Beaver in the summer of 1856, but did not move their families until the next year. On January 14, 1857, James bought 57 acres of land on what is known today as Martin's Bluff. It was not long before the widowed father of Edward's wife followed his daughter, bringing with him her young brother, and, before 1860, they were joined by Grace's two sisters and their families from Canada; Hannah, married to Whiskey Boyle (he later kept a saloon), and Katherine, married to Brian Don Mor Gallagher. The "Don" means brown, and the "Mor" either old or big.

The William O'Malley family left Mackinac for Beaver soon after the Mormon Exodus. William called himself a fish merchant in 1860, and in 1870 a hotel-keeper. With the family was two year old Sarah, remembered by many on the Island as Big Sal Dunlevy. An early visitor from Mackinac was James McCann, who then "decided to make Beaver Island his home," according to a biographical sketch in a local history.

32

He did not make the move, however, until somewhat later. Another Mackinac Islander who became a Beaver Islander was Richard Fitzsimmons. With him came his first wife, Margaret, whose fallen gravestone is to be seen in the Holy Cross Cemetery.

II

Because of proximity, the Mackinac Islanders were the first of the Irish to reach Beaver, but they were soon joined by others from Canada, New York and the coal mining regions of Pennsylvania. Almost all who came had a common background, for most had been fishermen in Ireland. Some families were from County Mayo, but by far the largest number had lived on either of the two islands off the coast of County Donegal, Rutland or Aranmore. One family was unique in that they had come from County Tyrone, a fact that earned them the nickname the Tyrone Gallaghers. Another similarity in background was that most of the early comers to Beaver were refugees from the Great Famine of 1845 to 1850. A few, like Black John Bonner, Patrick Kilty and the Luneys, had left before the Famine; some got away during the terrible years of hunger, but the greater number did not leave their native land until the early '50s, after the holocaust had run its course. These early '50s were the years of peak Irish emigration, for during the Famine itself many could find neither the means nor the physical energy needed to flee. Faced by starvation and its accompanying fevers, typhus and cholera, life had been lived from day to day and from hour to hour.

Contemporary quotations give some idea of conditions in Westport, home of the Martins, the McCanns and the O'Malleys, and of the horrors on Rutland, birthplace of Black Bonner and his relatives the Andy Roddys, and on Aranmore the home of the McCauleys, the Greenes, and most of the Beaver Island Gallaghers.

A Captain Perceval was Commissariat Officer at Westport. He told of a scene that took place on August 31, 1846. The people of the town had used every means they possessed to feed their hungry families; now they turned to the only source of help they knew, those in a position of authority. Captain Perceval wrote —

"A large and orderly body of people . . . marched in fours through Westport to Westport House and asked to see Lord Sligo. When Lord Sligo came out, someone cried, 'Kneel, Kneel' and the crowd dropped on its knees before him." The

Captain went on to say, "The state of Westport is indescribable; it is a nest of fever and vermin."

Lord Sligo had no help to give.

Things were no better in County Donegal. A relief worker visited the house of the Widow Cooney on Aranmore. The day before, the widow had buried her husband and one of her children, both dead from starvation. In her cabin, the visitor saw her five surviving children, "on their faces a gaunt, unmeaning, vacant stare ... Their lips were blanched and shrivelled from prolonged destitution." Nearby Rutland was equally stricken. George Hancock, member of the Society of Friends, reached there in 1847 as supercargo on the relief ship "Albert." Mr. Hancock was warned of the devastating fever that accompanied famine, and was told that he "endangered his life by standing in a crowd of about 200 wretched looking objects who were waiting for a distribution of food." How many future Beaver Islanders were wretched-looking objects in that starving crowd that surrounded the Quaker? Andy Roddy, later Captain Roddy, whose house, until a few years ago, still stood on the Sloptown Road, was then on Rutland, a boy of 13. His future wife, Catherine McBride, daughter of the local schoolmaster, was a child of seven that fatal summer of 1847. Was Edward Malloy able-bodied enough to row over from Aranmore for the distribution of food? We know he survived, for we have seen that he later followed his daughter Grace Martin from Mackinac to Beaver Island. When the "Albert" docked at Rutland, he was a man of 47 with four daughters whose ages ranged from eight through 18. How did Edward keep these girls alive through the years of desperation, girls who all later married and raised families on Beaver Island?

Not until the Famine had passed did many of its victims summon the energy to think about the future, and when they did, to a large number, that future seemed to lie across the sea. America was the land of promise, but none had yet heard of Beaver Island. Their first homes in America were in Canada, in the seaboard cities of the United States, or in the coal mining fields of Pennsylvania. It was in the Irish quarters of eastern cities that those who were the first to come to Beaver heard about the Michigan island that was to be their future home. Later, many came directly to the island, having been sent for by relatives.

Whether they came directly or stopped for a time in the East, it was typical that the journey was not made alone, nor as a single family unit. Rather, they came in large groups made up of people intricately related to each other and to those already in Michigan. It is impossible in this short account to cover the whole subject of Irish migration to Beaver Island. We will limit ourselves to a consideration of four typical groups. The first came from New York City, the second by way of Toronto, Canada, the third directly from Ireland in 1866, and the fourth, a group which left Aranmore in 1884 and was the last big party of Irish to reach Beaver Island.

III

It was probably Black John Bonner who was responsible for the coming of the New York party. Life was not easy for the Irish in New York City. Living in crowded and unsanitary tenements, set apart by a religion that was feared and despised by the majority of their new countrymen, they were greeted on every side by want ads and help wanted signs that ended with the words "No Irish Need Apply." In the spring of 1856 Bonner returned to New York to marry his fiancee Sophia Harkins, a native of Milltown, County Donegal. When he talked of an island set in a lake so large it seemed like the sea which surrounded their Irish homeland, where land was cheap and fish were plentiful, where there was unlimited work to be had cutting cordwood for steamships and ties for the growing railroad industry, it must have seemed to his New York friends that John was describing a Garden of Eden.

The Bonners were married in April, and in June the news of Strang's assassination reached them at their camp on Gull Island. As we have seen, they soon moved to Beaver. Word of the move must have been quickly sent back to New York to Sophia's relatives, her mother, now married to Philip Connolly with two young children by this second marriage, and to Sophia's younger brother and sister, Patrick and Bridget Harkins. Such a dramatic event as the assassination of a king would have been the subject of much discussion and endless speculation as friends gathered in the kitchens of the Irish Quarter of New York.

Among the Bonner and Connolly friends were Dan Malloy and his wife Fanny O'Donnell. Bonner had known Dan and Fanny when he had lived on Rutland and they on Aranmore. This was a couple who had been adults when the famine struck Ireland, and they were not married until 1852 when the years of hunger were over. By that time Dan was 37, Fanny 28, and it seems obvious that marriage for both had been delayed by the pressures of life in a country devastated by starvation and fever. When the terror had run its course, Dan and Fanny married and fled together from bitter memories and from the seemingly hopeless future offered by their native Ireland. Their first child was born in New York City in 1853.

Also in New York in 1856 were Dan's younger brother Jack and a sister, whose name we do not know. Brother Jack was a bachelor, but the sister had married an Irishman whom she had met in New York, William Gallagher from County Tyrone. Fanny also had a brother who visited them, Anthony O'Donnell. Anthony, nicknamed Salty, was an ocean-going sailor, the trade that had been followed by Black John before he settled down to fishing as a livelihood. The ships on which Salty sailed often brought him to New York where he was well known among the Irish from his native Aranmore.

Another relative of the family was a first cousin of the Malloy's brother-in-law William Gallagher. This cousin's name was James Peter Gallagher and to add to the confusion caused by the complexity of interwoven relationships, this Tyrone Gallagher's wife Bridget was the older sister of Grace Malloy whom we have seen married to Edward Martin on Mackinac Island. Does it help matters, or further confuse them, to point out that both Grace and Bridget were first cousins of Dan Malloy?

These complicated relationships are pertinent to the story of Irish immigration to Beaver Island. People do not live in a vacuum, they live among relatives and neighbors, and what one person does and says affects the lives of countless others. If we are to understand what brought the Irish to Beaver, how they came, and the life they created when they reached their island home, we have to at least attempt to understand what happened to them before they came, what memories they brought, and what relationships colored their actions.

By 1857 a few of the New Yorkers had decided to leave the crowded city for Bonner's fabulous island. The leader of the group was Dan Malloy, and making the journey with him in the spring of 1857 were his wife Fanny and their two children, his brother Jack, and his brother-in-law William Gallagher. William, later called Old Billie, was a newly made widower, for his wife, Dan's sister, had died with the birth of their first child. The child had also died and the bereaved husband decided to throw in his lot with his brother-in-law. When the party reached Beaver, landing at Cable's Bay, Dan and Fanny settled on the north side of Lake Geneserath. Here

Dan filed, on April 10, 1858, for 51 acres of land. There may have been others in this first group from New York, but of this we cannot be certain. What we do know is that, by the early '60s, several families from New York, relatives and friends of the Malloys and the Bonners, had reached Beaver Island — the Connollys, the Salty O'Donnells, the O'Briens and some of the Tyrone Gallaghers.

The Connollys left New York a year after Sophia and Black John had moved to Beaver. John had bought the present Bonner farm in 1857, but it was the next year before he got his house built, and in the meantime he and Sophia had lived in a deserted Mormon house on the site of the present Parish Hall. As soon as they were moved into their new home on the farm, the Bonners sent for Sophia's mother and stepfather. Together the two families built the house that now stands across the road from Bonners. This property was owned by Philip Connolly until it was sold to Feodor Protar.

Somewhat later than Sophia's family, the relatives of the Dan Malloy's began to leave New York for Beaver Island. Dan and Fanny were still living at Lake Geneserath when Fanny's brother Salty brought his family to Michigan in 1865. Salty had had a hard time persuading his wife to come to America, and an even harder time to leave New York for Michigan. He had married a widow with three children whose first husband had been another Anthony O'Donnell. Unrelated to Salty, this Anthony had been a saloon keeper and bailiff on Aranmore, and the family had been in better circumstances than most of their neighbors. After marriage in Ireland, Salty brought his wife and step-children to New York, and a year later a reluctant Hannah made a second move, this time to Beaver Island. Landing at Cable's Bay, the family went to the Dan Malloy home on Lake Geneserath. Hannah was bitterly disappointed with the primitive way of living she found at the home of her in-laws. Weary from the journey, she asked for a cup of tea, but this was a luxury the Malloys did not have. "At least," said Hannah, "we had tea in Ireland."

One family to come to Beaver Island from New York had an unusual history. Neither husband nor wife, Edward and Rosalie DeBriae, were Irish; they were French Canadians. By 1837, when the first of their seven children was born, they were living in New York City. It must have been there that they made friends among the Irish and thus heard of Beaver Island. By the time the 1860 census was taken we find them living on Green's Bay at McFadden's Point. Either while living in the Irish Quarter of New York, or soon after they reached Beaver Island, the family name was changed to O'Brien. The story told on the Island is that the Irish found De-Briae much too foreign a name for an Irish tongue to pronounce with ease. Someone remarked, "Oh hell, DeBriae is just French for O'Brien!," and O'Brien the family became. They are recorded as O'Brien in all census records, beginning in 1860, and as O'Brien or O'Brine in all land records. The first Mrs. O'Brien, however, must have felt some nostalgia for their French heritage, for on her gravestone in the Beaver Island cemetery she is, "Rosalie, wife of Edward DeBriae."

There is another story about a Beaver Islander who first lived in New York. Hannah Beag Rodgers, later Gallagher, and still later Boyle, is remembered for her small size, which earned her the Gaelic name of Beag, and for her large temper. She came from Ireland with her family while still a little girl. As the ship neared New York harbor, Hannah took her brogues on deck to give them a good shine. It was standard practice among immigrants, after the long, uncomfortable journey, to clean themselves and their clothes as best they could, in order to make as good an appearance as possible as they landed in their new home. When Hannah got the first shoe shined, she put it on the rail, and started to work on the second. There was a sudden lurch of the ship, and off went the shoe into the water, and an infuriated little girl, in a fit of temper, threw the second brogue after the first. Hannah landed without shoes.

Watching this fit of temper was an amused member of the crew, Black John Bonner. That night, after landing, John bought a pair of shoes, hunted up little Hannah, and gave them to her. Years later, in 1857 or '58, Hannah was on another ship, this time with her young husband and a two year old baby. As the boat pulled up to the dock on Beaver Island, Hannah cried to her husband, "John, John, there on the dock is the man who gave me the shoes!"

The last of the Malloy relatives to leave New York for Beaver Island was the family of James Peter Gallagher. James was a cousin of Old Billie, Dan's brother-in-law, and his wife Bridget, nee Malloy, was a cousin of Dan himself. We will now call James Peter by his Beaver Island name, Big Gallagher. This family stayed in New York until 1871. Big Gallagher was 53 years old when they made the move to Beaver and with them came their 16 year old son William, known by all on Beaver Island as Bowery Bill. This nickname, coined by Black John Bonner because Bowery had been born in New York City, is applied to his descendants; to this day they are all Bowerys.

When Big Gallagher got to Beaver Island he was among relatives. Not only was cousin Old Billie there, but also Old Billie's brother, Mike Mahal Ruah, who had reached Beaver Island by way of Canada. I was told on the Island that Mahal means, "myself," but the Gaelic Dictionary gives — "Mahail — doggish, fierce, surly." Ruah, of course, means red. Mike's wife, Susan Mooney, also had Beaver Island relatives, for her sister Mary was married to Pat Malloy. Although the two Malloys, Pat and Dan, were both from Aranmore, they were unrelated. For Big Gallagher's wife Bridget, the arrival on Beaver must have been a joyful occasion, for not only her cousin Dan Malloy was there to greet her, but also others even closer. Here she was reunited with her 68 year old widowed father Edward, and her three sisters, Grace, married to Edward Martin, Katherine, to Brian Gallagher, and Hannah, to Whiskey Boyle. The family had not been together for many years, for while Bridget had been with her husband and children in New York City, the others had been living in Canada and Michigan.

IV

A party that may have reached the Island even before the one from New York City came from Toronto, Canada. There is a story about how this group heard of Beaver Island which recurs, in slightly different forms, again and again among the Island's traditional story tellers. The hero was Charles O'Donnell, called Charlie Strack. Strac in Gaelic means strike or beat, but how he earned this nickname we do not know. Charlie and his wife, Grace Gillespie, lived in Toronto where there was a large concentration of Irish immigrants, many, like Charlie and Grace, from Aranmore. Working on a railroad construction gang, Charlie was entrusted by the gang foreman with money for the payroll. So large a sum of cash was too great a temptation for O'Donnell, and he absconded with the money, fleeing across the border. This was a common offense of the times. In the Catholic newspaper, "The Pilot," a priest in New York State wrote that in his area, within a few months, no fewer than six sub-contractors, two of whom were Catholic Irishmen, had stolen company payrolls, and departed for parts unknown.

Safely in Detroit, Charlie got a job with a lighthouse construction crew which was being sent to Beaver Island. One version of the tale is that he wrote to his wife in Toronto telling, in glowing terms, of the island he had found, an island remarkably like Aranmore, and full of empty houses which had been abandoned by the Mormons and could be had by any who simply walked into them. The letter urged Grace to tell their Irish friends about this great opportunity, and to persuade them to bring her and the baby, and join him on Beaver Island. Another informant has told me that, instead of sending a letter, Charlie saved what he earned working on the lighthouse, and then returned to Toronto, paid back the money he had stolen, sang the praises of Beaver Island, and came back with those whom he had persuaded with his convincing Irish tongue. The latter story appears to be the more probable. It would seem that Charlie in person would have been more persuasive than a letter, particularly as, in 1880, he told the census taker that neither his wife nor he himself could read or write.

Charlie's story of empty Mormon houses means that the construction gang in which he worked must have been the one sent to build the Harbor Light, for the Lighthouse at the Head dates back to 1851, during the years of Mormon supremacy.

We cannot be sure of the identity of all the families coming from Toronto, but of some we are certain. Prominent among them were Conn McCauley and his pregnant wife Mary Gallagher, and the Sylvester McDonough family consisting of Sylvester, his wife Ellen Corey, and their two children. The McDonoughs had married in Canada, and, unlike most Beaver Islanders, they were from neither County Donegal nor County Mayo. Sylvester, who was never called anything but Vesty (even his land records use that shortened version of his name), was from Galway, his wife from County Clare. Traveling with the McDonoughs was Vesty's sister, Big Mary. Big Mary came honestly by her name. She was six feet tall and so strong that when the boats were too high on the beach, making it difficult or impossible for the men to get them back into the water, Big Mary was called on for help, and putting her shoulder to the task, this Amazon soon had the boat afloat.

To continue with our roster, also in the party were: the widower Dan McCauley (unrelated to Conn although both were from Aranmore), Dan's son Edward, and his nephew Dan Boyle. With Dan Boyle was his pregnant wife Katherine

Gallagher, better known on Beaver Island as Katcheline Mor. Another Daniel in the group was Daniel Gallagher, a cousin of Conn McCauley's called Danny Don Mor, and with him were his wife and baby. The party also included one bachelor, 21 year old Hugh Connaghan. Our list of those known to have come from Toronto at this time ends with the Anthony O'Donnell family. This Anthony, although also from Aranmore, was not related to the Anthony called Salty who came to Beaver Island by way of New York City. This Toronto Anthony may have heard of Beaver Island not only from Charlie Strack, but also from the brother of his wife, Sophia, for this brother was Johnny the Rat, fishing partner of Black John Bonner.

Although the Anthony O'Donnells end our list of families we are certain were in the Toronto party, there is another which, in all probability, we can safely assume made the trip with them, Dominick Gallagher, his wife Mary Greene, and their baby daughter Bridget. Big Dominick, to call him by his Beaver Island name, was the brother of Conn McCauley's wife Mary. The Gallaghers' oldest child had been born in Canada in 1855, but the next daughter, Katherine, was born on Beaver Island in 1857. Because of Big Dominick's relationship to Conn McCauley's wife, and the date and place of daughter Katherine's birth, I think we can safely assume that this family was passengers on the boat from Toronto.

A granddaughter of Conn McCauley, in recalling what was told by her grandmother, said that the Toronto party got to Beaver Island "just three months after the Mormons had left," and that they moved into empty Mormon houses which "were filled with broken furniture and crockery. What the Mormons could not take with them, they destroyed." I am sure that this was a statement of what grandmother Mary McCauley believed to be true, but to a disinterested observer many years after the fact, it seems more likely that the destruction was done by those who had driven the Mormons from their homes. Another story one hears, and which seems aprocryphal, is that "the Dan Boyles moved into a house where the hearth was still warm and the cow still in the barn."

Both the "three months after the Mormons left," and the "warm hearth" are hard to reconcile with information given later to the census taker by both Vesty McDonough and Dan McCauley. In 1900, for the first time, the census asked the year of admission to the United States, and both 83 year old

Dan and 70 year old Vesty gave the date as 1857, a year after the Mormon Exodus. Dan McCauley, unlike most of the early Beaver Islanders, was a man of some education, for he had been a schoolmaster in Ireland. The memory of facts which were not recorded at the time they occurred tends to be faulty, but it seems unlikely that schoolmaster Dan would have mistaken the year of so important an event in his life as his entrance into the United States. Another argument for the 1857 date is that with Dan Boyle when he reached the island was his wife Katcheline Mor, and the Boyle family Bible recorded them as not having been married until 1857. The information in the census, plus Dan's marriage date, must be weighed against an oral tradition for 1856 which is very strong among descendants of some of the Toronto families. We will have to leave this discrepancy unresolved.

Uncertainty may exist about the exact year in which the journey was made from Canada, but we do know how it was made. The group chartered a boat in Toronto, and, piloted by Conn McCauley, in this way reached the island of Charlie Strack's glowing description. Of just where they first lived when they got to Beaver, we are not certain. Family tradition says that the Dan Boyle's first home was on French Bay, but it was on Green's Bay that he acquired his first land, homesteading its southern end. On this bay twelve families at one time made their homes, including not only Dan Boyle and Big Dominick, but also the O'Briens from New York, and, later, the Greenes from Ireland. Within a year the McDonoughs, the Dan McCauleys and Hugh Connaghan were living on Big Sand Bay where another group of families had settled, setting up a community which eventually included two stores, a school house and a sawmill.

Like those who came from New York, the Toronto group was soon joined by relatives. Vesty's wife Ellen had left a married sister in Toronto, Bridget Burns. Bridget's husband was drowned, and she came with her three children to Beaver Island, where she soon married the widower Dan McCauley. Dan had been the first of his family to have left Ireland, and while still in Canada he had sent back passage money for two nephews. First to come was his sister's son, the Dan Boyle who was a member of the Toronto party. The circumstances of Dan's marriage are still a cherished memory among his descendants. After he had been in Canada a year or two, he went back to Donegal to marry his sweetheart, only to find that she had left Ireland and come to America with her parents. Undiscouraged, the would-be bridegroom came back, found her in Canada, and they were married in time to come to Beaver Island with his uncle.

The second nephew for whom Dan McCauley had sent was his brother's son, Pete McCauley. Pete and his friend Hugh Connaghan had left Aranmore together, joining Uncle Dan in Canada, and we have seen that Hugh was a member of the Toronto party to Beaver Island. Pete was not, for he had already left Canada for the United States when the party was organized. His mother and stepfather, Mary and Tom H. Boyle were in Pennsylvania, and Pete had gone to join them in Johnstown where he worked in the rolling mills. To the best of our knowledge, Pete's stepfather was unrelated to his cousin. After news came that their friends and relatives were on Beaver Island, the entire family decided to go to Michigan. The family included — Pete, his mother and stepfather, a half brother and a half sister who were still children, and two stepbrothers, grown men like Pete himself. Pete was the McCauley for whom McCauley's Bay was named, for in later years it was there he bought land and established his fishing business.

The other McCauley of the Toronto party, Conn, (unrelated as we have said to Dan and Pete) was one of five brothers, all of whom eventually made their homes on Beaver. At the time the group left Canada, Conn was the only one of the five to have emigrated from Ireland. His brother James was the next of the family to leave, joining Conn in 1860. Four years later the two became brothers-in-law as well as brothers, for James married Conn's wife's sister.

V

In the years between 1856, when the first Mackinac Irish moved in to fill the vacuum left by the departure of the Mormons, and the beginning years of the decade of the 1860s, Beaver Island was a place for the gathering of the clans. Ties of blood and friendship were strong, ties that had been forged in the close knit island communities of their Irish homeland. The new island reminded them of the old; in family after family the words have come down, "it was like Ireland." But, unlike Ireland, here in America there was not only good fishing, but also land so cheap even a poor man could buy, a much prized boon that had been denied them in their native land. Word first went to friends and relatives already in the United States and Canada, but it was not long before letters were sent to Ireland. Later some Beaver Islanders made trips back to Aranmore and Rutland, and there extolled the wonders of their American island. The virtues of Beaver lost nothing in the telling, and the stories of the visitors took deep root in the islands off the coast of Donegal. As late as 1965 I found that the words "Beaver Island" had a magical effect in Aranmore, opening doors and producing smiles and a hearty welcome. Old men told me tales of an island they had never seen, but that they firmly believed to be a veritable Garden of Eden.

Some of the Irish families on Beaver soon saved enough money to help bring over those who had been left behind, and letters crossed the Atlantic urging relatives to come to America. There must have been many gatherings around peat fires in the cabins of Aranmore where letters were read and re-read, plans made, unmade, and then made again. Winter passage was impractical, but with the coming of spring, plans matured and groups of friends and relatives resolved to undertake the arduous trip together.

One such group arrived at Cable's Bay a few days before July 4, 1866. They had come directly from Aranmore to Beaver Island. At that time the principal town in their area of Donegal was located on the now deserted Island of Rutland. I was told on Aranmore that many of the emigrants boarded the ships that brought them across the Atlantic in their home town of Rutland, walking up the gang plank carrying over their shoulders sacks of oatmeal to be used for food during the six weeks' passage. The make-up of the 1866 group was complicated but instructive. If we can untangle its relationships, we will have gone a long way toward understanding how the Irish community of Beaver Island developed.

In the group were three people, well along in years, who would be joining children already on Beaver. These were the widow, Kitty Gallagher, age 64, Frank McCauley, 68, and his wife Ann, 65. Kitty's son, Big Dominick, and her daughter, Conn McCauley's wife Mary, had been in the Toronto party of 1856 or '57, and another daughter Alice was also on Beaver, having, as we have seen, married Conn's brother James in 1864. In these three families Kitty had eleven grandchildren waiting to welcome a grandmother they had never seen. With her came her youngest son, Owen, Big Owen on Beaver Island. The records and the memories of Beaver Islanders are confused as to just when a third son arrived. Like his brothers, this man carried the nickname Big; he was Big Neal. It may have seemed to Kitty, age 64, as she said goodbye to all she had known and loved, that she had only a short time left to spend with her children and grandchildren. In reality there lay ahead of her almost forty useful and happy years on Beaver Island, for she died in 1905, age 103, her life having spanned two worlds and more than a century.

Frank and Ann McCauley had even more children awaiting them than Kitty. This couple were the parents of five sons, all of whom sooner or later reached Beaver Island. We have seen that sons Conn and James were there before 1866. The McCauley's grandaughter, Mel Big Owen, has told me, "John was the last of the family to come, and he brought his mother and father with him." If correct, this means that Paddy Baca and Owen were already on Beaver awaiting their parents' arrival, but Owen, in 1900, gave the date of his immigration as 1866, which would have made him a member of the party we are now considering. Definitely with her father and mother was twenty year old daughter Hannah, and a friendship, cemented by the trip, soon became more serious, for she married Kitty's son Big Owen, and became the mother of Mel Big Owen, to whom we are indebted for much information.

Not only Kitty's Big Dominick, but also his wife Mary Greene had relatives in the Aranmore party. These were two first cousins, both named Daniel Greene, who were also first cousins to each other. On Beaver Island they were called White Dan and Red Dan, and, as if we needed a further complication, I have been told that "Red Dan's mother was first cousin to Big Dominick Gallagher." There were other relatives of the two Dan's on Beaver besides Mary. Neil Greene, either Mary's brother or her uncle (family memory is very unclear on this), had, in 1865, started proceedings under the Homestead Law to acquire land at Green's Lake, and her bachelor brother, Pete, called Bogy, was on the Island by 1864.

Both Dans were unmarried when they emigrated; Red Dan was 28, White Dan 25. They saved enough money to send, the next year, for White Dan's two younger sisters, Bridget and Sarah, whom they had left living unhappily in a home with a bitterly resented stepmother. Before 1880 their father and his second family also came to Beaver, but by that time, well-established in their own homes, the sisters seem to have been free of any earlier ill-will. The cousins, White Dan and Red Dan, lived at Green's Bay, and about 1870, they were joined there by Red Dan's brother Owen, his wife and four children. "All the Greenes lived on Green's Bay," I have been told; but in the '70s that settlement was abandoned and the Greene families moved to farms in the interior. Again the Greenes gave their name to where they lived; the area along the Kings Highway below the Sloptown Road we still call Greenetown.

At the time the party of Gallaghers, McCauleys, and Greenes left Aranmore, Mary Ellen Roddy lived on nearby Rutland. Contradictory records make her exact age at this time uncertain, but she was in her early twenties. In 1900 she gave her immigration date as 1866, so presumably she made the trans-Atlantic passage with her friends and neighbors. Like the others, Mary Ellen was coming to join relatives, for she was a niece of Black John Bonner, and her brother, Andy Roddy, had already moved his family from Rutland to Beaver Island. When Andy had first come to America he had left his wife, Catherine McBride, daughter of the Rutland schoolmaster, behind with their baby son. Like many Irish, his first job had been in the coal mines of Pennsylvania, but Roddy disliked the mines and came farther west. For some time he worked on trading vessels sailing out of Chicago, and when they stopped at Beaver Island he visited with his uncle, Black John. Attracted by Beaver Island, Andy gave up the idea of living in Chicago, sold for a song the land he owned where the Marine Hospital now stands in that city, sent for Catherine and baby Andrew, and settled his family in the two-story log house which, until a few years ago, stood on the Sloptown Road. Captain Roddy was long remembered on Beaver Island for his zest for life, and for his beautiful singing voice. "It was as good as John McCormack's," I have been told again and again.

Within a year after getting to Beaver, Mary Ellen married a Rutland Gallagher named Barney, son of Paddy Grey Gallagher. This family, "not related to any other Gallaghers on the Island," to quote my informants, like Roddy, went first to the coal fields of Pennsylvania, where both Barney and his father worked in the mines. Paddy disliked the life of a coal miner and the family returned to New York. It must have been there that they heard of Beaver and, although they were on the island by 1858, they probably came as a single family, for I have never been told that they were part of the New York party of 1857.

When Mary Ellen, the McCauleys, the Gallaghers, and the Greenes landed at Cable's Bay a few days before July fourth, the Islanders were already celebrating the national birthday of their adopted country, celebrating as only the

46

Irish can. I have been told the festivities began a few days before the fourth and lasted for a few days after it had passed. It does not take much imagination to comprehend what this day meant to these new Americans, natives as they were of Ireland, and resentful of the long tyranny they had endured under the hated English rule. This was the day of freedom, the day of British defeat, and the Island was celebrating with such abandon that at least one of the arriving newcomers thought she had reached a country of madmen. As long as she lived, Hannah McCauley, later Hannah Big Owen, told of that momentous arrival, and how, not knowing this was a special day, she thought that she was witnessing the ordinary behavior of those living in the country that was to be her home. On that day she believed that in America all men, at all times, acted like madmen.

VI

The people in the last big party to come from Ireland to Beaver had no such dramatic introduction to their new life. They arrived in 1884, and, like those in the 1866 group, had relatives awaiting them. The largest family was that of Daniel Peter Gallagher, pronounced Dōn Father. Gallagher was 48 years old in 1884, and had already been twice in America, and had twice returned to his home on the little island of Inniscara, which lies between larger Aranmore and the mainland. On this third trip he brought his wife and seven children, and, after landing in Boston, they came directly to Beaver Island where Don Father had decided to make his permanent home. When the family reached the island they were among many relatives; Frank McCauley was Don Father's uncle, being his mother's brother, and this made him a first cousin to the five McCauley brothers. One of the brothers, Paddy Baca, is remembered by Don Father's daughter Mary, a child of eight at the time, as having been a member of the 1884 party. This could not have been Paddy's first trip to Beaver; he must have been returning after a visit to Aranmore. On his father's side Don Father had three first cousins already on the island, Manus Gallagher and the two brothers, Cornelius and Cundy Gallagher. Cornelius had left Ireland in 1851, and, on reaching America, went to the Pennsylvania coal mines to work. The family lived in Pennsylvania for ten years. When they came to Beaver it was planned as a visit, but "it was so like Ireland," the family stayed.

Little Mary Gallagher, later Mary Early, remembered that Belle Daugherty and her family made the trip from Ireland with this group. Another who traveled with them was the widowed Nancy Gallagher and her five year old daughter Nora. Nancy was better known on Beaver Island as Nanj Og, meaning young Nancy, and two years after her arrival she married Darkey Mike O'Donnell, the man who gave his nickname to the road on which he and Nanj Og lived. Other than these names we have listed, the make-up of the 1884 party is unknown, although there were probably some whom eight year old Mary failed to remember.

VII

As we have seen, many Beaver Islanders reached the island in large groups, but some came as single families and others as individuals. After the arrival of those from Mackinac, most of the Irish came in answer to letters and made the trip with passage money sent by relatives. This is a recurring theme in the recollections of older people. "Anthony O'Donnell sent for Paid een Og and his family; Paid een Og was his wife's brother." "John McCauley sent for his bride, Katcheline Og." "Red Dan sent the money for his brother Owen."

One story is so distinctive and of such human interest, I cannot resist telling it here. When the Labbly O'Donnells left Ireland in the 1860s, they had only enough money for their own passage, and their children were left in the care of relatives. This man is always referred to as Labbly, a word that seems to be a corruption of the Gaelic Labhar, meaning noisy or loquacious. The memory of his given name is lost even among his descendants, but from the records I think it must have been Daniel. Times were hard; several years passed and Labbly and Mary had great difficulty saving enough money to send for their children. Finally, in desperation over their divided family, they had an original idea; they would raffle a sheep to get the money. Friends and neighbors co-operated; tickets were sold; enough money was raised and the children were sent for. When Dennis Cull, a bachelor, bought his ticket for the raffle, he said, "When your daughter Mary gets here, I am going to marry her." This was a great joke; Mary was 12 or 13 and Dennis 33, but in the end, he made good his promise. Five years later, Dennis and Mary were married.

This has been an incomplete account of the emigration of the Beaver Island Irish and of how they reached an island that seemed to them a fulfillment of their hopes. "It was like Ireland," they said. But it was not Ireland, and the future lay before them. As they built their new community, how much did they keep of the values and the customs of the life they had left, and what adaptations did they make to a new and different world? That is another story.

F. PROTAR:
THE HEAVEN-SENT FRIEND

by Antje Price

On the western shore of Beaver Island, atop Bonner's Bluff, near the old logging railroad right-of-way, one unexpectedly comes upon two monuments. On the right, a plain wooden cross stands at the site where Engineer Chase died when his logging engine overturned in 1913. Across the road, at the edge of the forest, a field-stone enclosure with a white wooden gate marks Protar's grave. A bronze plaque bears his likeness and the words "To our heaven-sent friend . . . who never failed us / in imperishable gratitude and admiration / his people of Beaver Island."

Protar, an émigré from Russia by way of Germany, settled on Beaver Island in 1893, after successful careers in the theater and newspaper publishing; until the end of his life in 1925 he served as family doctor to the islanders, charging not a penny for his ministrations. Now, more than fifty years after his death, his memory is still very much alive on the Island; the feelings of gratitude and admiration have not perished and the air of mystery about his origin and motivation has not been dispelled.

The memory of this remarkable and enigmatic man has been preserved mainly in stories and legends. Some of these have been published, including a memoir by Paul Kersch, Protar's younger friend and protégé, first printed in *Michigan History* in 1939 and now available as the booklet "The Saint of Beaver Island" from the Beaver Island Historical Society; some newspaper articles (in the Detroit Public Library); a chapter in *The Saga of Beaver Island* by Kenny and Cronyn (1958) based on these; and *Russian Doctor in Paradise* by A. Ludwick (1962). All of these relied heavily on conjecture, often stated as fact, especially in regard to Protar's pre-island life, which he was unwilling to discuss and about which he apparently tolerated some assumptions. Many of the accounts, however, are contradictory to historical and geographic fact, as well as to known circumstances of island life, presenting Protar as a fictional hero rather than the actual but no less heroic man he really was.

In recent years, more direct sources have become available, permitting the real Protar to emerge. These include his medical ledger for the years 1915-25; brief diary notations for 22 of his 31 island years; his Bible, a prayer-book (used 1857-66) and a book of philosophical readings, compiled by Tolstoy (used 1907-25), with many markings and marginal notes, some of them dated, as well as other books from his library with notes and dates. All but the medical ledger were written in German. Most of these items are in the possession of the Beaver Island Historical Society and were made available to the author by the curator of its museum, A. J. Roy; a few are in private hands and were also generously shared.

This account is based on the newer findings, including translations of the diaries and notes (by the author), library research stimulated by them, on-site study of material in Rock Island, Illinois, and accounts obtained from tapes or by interview from persons who had known Protar, both in Rock Island and on Beaver Island. It summarizes the current status of information from these sources, which are being pursued further for a larger biographic study. Since the work is far from complete, it can only be presented here as a preliminary account, serving to introduce the new data and to give added perspective to Protar's island life. No information will be stated as fact unless it has been verified by two or more independent sources; unverified items will be stated as such, vague hints and conjectures will be labelled, if given at all. Discussion and refutation of earlier reports will not be attempted here.

The newer findings are at some variance with old accounts and tend to dispute many of the legends about Protar. However, they do not diminish him, but rather add to his stature and begin to define him as a whole man.

Any study of Protar and attempts to know him well are impeded by the fact that he did not want to be known and went to considerable lengths to conceal his identity and past, including adoption of a pseudonym. Those closest to him agree that he was a haunted, hunted, desperate man, but only one person knew the whole story of his troubles, his old friend Dr. Carl Bernhardi, who took it to his grave. Many considerations argue against pursuit by the Russian government, as legend would have it, among them the fact that he changed his name *to* a Russian-sounding one. Exile from his homeland appears to have been self-imposed. The nature of his nemesis is not revealed by the diaries; however, his notes show that he was his own most ardent and tireless persecutor and that his flight and search for refuge were largely within and against himself.

"Protar" is an anagram of his real family name, Parrot. His given name was probably Friedrich (Frederick), like his father's and grandfather's, but this has not yet been verified. His assumed first name is given variously as Feodor (Kersch) or Fedor (Rock Island, his Will) and corruptions of these (as on his tomb: Feodora); he usually signed his name "F. Protar." Through his family name, given dates and notes in his Bible, prayer book and other volumes, it has been established that he came from a family of eminent professors at the University of Dorpat (now Tartu) in the then Russian Baltic province of Livonia.

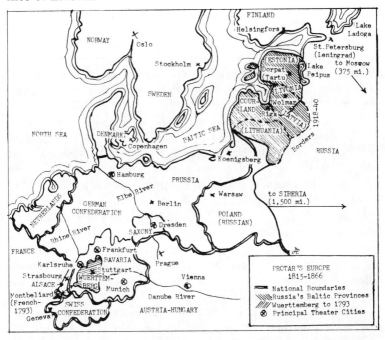

Livonia, at the south-east end of the Baltic, south of Finland, was Christianized in the 14th century by German knights returning from the Crusades, and their descendants formed the ruling, land-holding and urban elite — Dorpat was a Hanseatic city — while the native Esths and Letts constituted the peasant and laboring class. It came under Polish and then Swedish rule, at which time Protestantism was introduced and the university at Dorpat was originally founded. It was acquired by Russia through treaty in 1721, but retained considerable autonomy, including language (German), customs, and property law (Swedish). Russianization was not attempted until after 1880. The Estonian drive for identity and independence began as a student movement at the Dorpat university and gained momentum through a series of song festivals — it was said that Estonia sang its way to freedom. Independence was finally achieved in 1918, but lost again to Russia in 1940. It is felt that this brief historical overview is necessary to an understanding of Protar's "Russian" background.

Protar's grandfather, (Georg) Friedrich von Parrot (1767-1852), of French descent, was born in Montbeliard, Upper Alsace, then part of the German Duchy of Wuerttemberg, but annexed by France in 1793. He was the scion of an extraordinary family of scholars, physicians and scientists, traceable through their writings and biographies from 1733 to the present in Alsace, southwestern Germany and Paris. Eleven of the eighteen men located to date were physicians. After completing studies in economics, mathematics and physics at Stuttgart in 1788, Friedrich Parrot took a position as tutor in France and then as a teacher of mathematics in Germany. During this time he married a young woman from Geneva whose family had connections with the Russian court. They had two sons, but his wife died in 1794. The next year Parrot emigrated, with the two boys, to Livonia to serve as tutor on a country estate. He was married again in 1797, to the daughter of a local German aristocrat; it is not known if they had children. The university at Dorpat was being refounded at the instigation of Czar Paul I, who had known Parrot at Stuttgart, according to Protar's notes. Parrot was appointed professor of physics at the new university, and as leader of the faculty, became its first chancellor (1802-18). Through close friendship with Czar Alexander I (1801-25) and association with Nikolai I (1825-55), he had considerable influence on educational policy and peasant rights in Livonia, as well as higher education in Russia. From 1826 until his retirement in 1840, Parrot was a member of the Imperial Academy of Science in St. Petersburg. Among his many contributions were findings which formed the basis for later theories of osmosis (Dutrochet) and dialysis (Graham), as well as galvanic electricity (Faraday). His many practical inventions included a lightning rod. In many ways he could be called the Russian Franklin, or rather, Franklin the American Parrot.

The title "von," equivalent to French "de" and British "Sir," was not hereditary in the Parrot family and did not denote nobility. Those Parrots who carried it had it conferred upon them as an honor in recognition of service or achievement by their respective sovereigns. They include Georg Friedrich von Parrot, his two older brothers and his eldest son Friedrich, Protar's father, who was honored with a knighthood for his scientific explorations, specifically the first ascent of Mt. Ararat in 1829.

(Johann Jacob) Friedrich (Wilhelm) von Parrot (1791-1841), sometimes called "the younger," Protar's father, was no less eminent a scientist than his father, eulogized at his untimely death by Alexander von Humboldt himself. He interrupted his medical studies in Dorpat for a scientific expedition to the Caucasus in 1811 and to serve as military physician for the Russian Army in 1812 and 1815. In 1820 he was appointed professor of physiology and pathology at Dorpat university. In 1826 he took over his father's chair in physics, which had been his first love. Like his father he had two wives: in 1821 he married his first cousin, who died in 1825, leaving a small daughter; he married Emilie Krause, daughter of a Dorpat professor, after returning from the Ararat expedition (1830), and she bore three sons, one of whom was Protar. After an expedition to the North Cape to study pendulums and magnetism in 1837, Parrot became seriously ill and died three years later. He had been the original owner of Protar's little prayer book, and the notes in this led to the discovery of the lineage. Protar's mother lived until 1866. Parrot's younger brother, Protar's uncle Wilhelm, was a Protestant pastor in Livonia. It should be noted from the above that Protar was neither of Russian nor of noble ancestry, but that his family possessed and passed on in large degree an aristocracy of the mind and a nobility of public service.

Protar was born on August 22 in Dorpat, probably in 1837. All American sources give this year and all age reckonings are from this date. In his Bible the final numeral of the year is not entirely clear — it is neither a proper German 1 nor a 7, simply a vertical line. It has been interpreted as a 1, but this has no verification from other sources, and since all indications suggest the later date, 1837 is being accepted at this time. He was confirmed in the Lutheran faith in 1849 in Wolmar, a town in southwest Livonia between Dorpat and Riga, which may have been his uncle's parish. It is said that he attended the university at Dorpat, and this appears likely, probably on a medical or scientific course of study, which was, however, apparently not completed, since he did not make use of it to launch a professional career, following in the albeit formidable footsteps of his father and grandfather. There is no doubt that he was well educated and that he had ability, he chose lines of work in which he could use his education but which did not require the credential of an academic degree: the theater and journalism.

According to his notes, he was in and out of Dresden, in Saxony, and Dorpat between 1857 and 1864. There was no university in Dresden, but a very active theater, and it is assumed that he began his stage career there. Fanny Janauscheck, the great Bohemian tragedienne with whom he later toured, was in Dresden during that time. A visit to St. Petersburg is noted, which may coincide with a command performance he reportedly gave for the Czar. Between 1863 and 1866 a theatrical tour also brought him to Koenigsberg in Prussia for his first meeting with Carl Bernhardi, a medical student at that time.

Protar was reportedly Mme. Janauscheck's stage manager, as well as an actor in her troupe. She first came to America in the fall of 1867 to perform in German theaters, principally in New York, and returned to Europe in 1869. It is not known whether Protar was with her. She came again to New York a few years later and made her English-speaking debut under Augustin Daly's management in 1871 or 1873, and then remained in America, performing both in German and English in New York and on tour, until her retirement in 1896. Protar's notes are completely blank for the period 1867-73, with one significant exception: in his Bible, the notation "Louise — marriage October 12, 1868" and then "died June 22, 1872," referring to I Peter 5, 6-11 for his own guidance and consolation. It is not known where he was during this period, but one slim hint that he may have been in America is the fact that, though the Bible notation is written in German, it is not written in German script, as only two years earlier the record of his mother's death had been. It seems unlikely that he would have been on tour during that time. He reportedly edited the German edition of *Leslie's Magazine* in New York, but this has not been verified and the dates have not been established. At any rate, he apparently returned to the stage after his wife's death, because from 1874 to 1878 he bought books of plays in New York and St. Louis (1875), annotating some for performance or reading. December 1874 marks the first appearance of the name "F. Protar." There is another blank period from 1878 to 1882, but also fragments of play scripts that cannot be dated.

Protar next emerges in Rock Island, Illinois, in July 1882 as editor and publisher of the *Volks Zeitung,* a German-language newspaper. Rock Island was a pleasant river town, part of the "Quad Cities" contiguous with East Moline, Moline (in Illinois) and Davenport just across the Mississippi in Iowa. They boasted large German and Swedish populations with their own churches, schools, social clubs, theaters and newspapers. The Davenport City Theater hosted an annual season of German drama from September through May, booking performances by touring stage companies. The *Volks Zeitung* had been founded in 1875 as a bi-weekly paper, but its owner went bankrupt in 1882. It was sold in March of that year and resold some time before July 1 to Protar. How and when Protar came to Rock Island and why he decided to stay there has not been established, but two coincidences should be considered, beside the fact that the paper was available and Protar was an experienced editor: Dr. Carl Bernhardi had settled in Rock Island and opened a practice there in 1869, and Fanny Janauscheck's troupe appeared at the Davenport theater for two performances in May 1882. At any rate, Protar took over the newspaper, and editing it in a "liberal and progressive spirit" built it into a successful enterprise. According to Val Peters, a later publisher, "Under the direction of Mr. F. Protar the newspaper took a decided upward turn. It was independent in politics and had a great influence in city and countryside." The offices of the paper, as well as Protar's lodgings, were on the second floor of the Buford block, near Rock Island's market square.

During the summer Protar traveled on the Great Lakes, and on one such trip a storm reportedly brought his ship into Beaver harbor. He fell in love with the island and continued to vacation there. In the summer of 1892, while staying at the Gibson House (now the museum), he went along on a trip to the Bonner farm to get fresh vegetables for the hotel kitchen. Noticing a vacant log house across the road, he asked about its availability. It had been built in 1857, reportedly from

Mormon logo, by Bonner relatives, the Connellys who had moved to Escanaba about 1887. The sale was negotiated and Protar acquired the house, barn and 200 acres of forest and farmland in late April, 1893. On April 1, he had sold the newspaper to a consortium of German citizens of Rock Island and Moline. Paul Kersch, who first appeared in Rock Island in 1891 as a printer on the paper, became business manager, working with a succession of editors until he gave it up in 1897. The paper passed through several hands and finally went bankrupt again. "The guiding spirit was missing," wrote Val Peters, who took it over in 1901; "there just wasn't any Protar any more." The *Volks Zeitung* revived under Peters, but he moved it to Nebraska in 1910, where it survived as the *Omaha Tribune* until at least 1921.

Protar moved to Beaver Island in April 1893 and as far as is known, never left it again. Apparently he had a considerable sum of money at his disposal from the sale of the newspaper, which he husbanded carefully to support his island existence. More than 25 years later he sold some timber rights to bolster his finances. At first he planned to farm his acres, but when the man hired to help him turned out to be incompetent and arrogant besides, Protar sent him packing and gave the logs, already cut to house him and his family, to the McCaffertys down the road. Apparently deciding that farming was too large and risky an undertaking for him, he abandoned the venture and presumably cast about for the best way to contribute his skills to the island community.

His most useful role revealed itself to be that of family doctor and pharmacist to the island population. For thirty years he dispensed the necessary medications, as well as care and advice, to all who needed it, without any charge. He practiced palliative medicine to alleviate common and chronic ailments, such as respiratory infections, digestive upsets, worms, childhood illnesses, arthritis and neuralgia, surface wounds and skin rashes, rather than curative or acute care for major ailments. He did not handle narcotic or prescription drugs, vaccinations, serious wounds and diseases, or childbirths. Cases beyond his capability were sent to Charlevoix or a physician was called over from the mainland, if there was not a physician on the island at the time. From 1900 to 1911 there were five doctors residing on the island at various times; there was one in 1914, and from December 1923 the state supported a physician at St. James, Dr. Palmer. Protar did not call himself "Doctor," nor would he tolerate that designation from others.

Protar reportedly obtained his medicines from the wholesale drug firm of Hartz and Bahnsen in Rock Island; one of the proprietors, Ben Hartz Sr., appears to have been a friend, who also visited him on Beaver Island. According to his ledger, the drugs most frequently given, year around, were cough medicine — sometimes by the quart for whole families — and bicarbonate of soda, along with fever reducers, worm medication, tonics, and analgesics. Drugs came in the form of pills, drops, powders, tinctures, washes, ointments, dressings, plasters, liniments, gargles and snuff. Not listed in the ledger, but certainly given as well, were sitting-up-with, comforting, advising and admonishing.

He treated Gallaghers and McDonoughs, listed by their island names spelled phonetically, Greens and Martins, La-Frenieres and Pelletiers, Ricksgers and Schmidts, Indians, Priests and Sisters, Dr. Bernhardi and himself ("ego"), and occasionally a horse or calf. Legend has it that he drove (with horse and buggy or cutter) to his patients, but stories also indicate that his patients came to him, which appears to have been the more frequent mode, in keeping with his palliative practice. Some ledger entries list various treatments for several family members on the same day, as if he were taking care of them all while he was there; the diary entries indicate days when roads were impassable and "horses not out of stable." A tabulation of his ledger entries shows that he saw patients on the average 11 days per month, averaging 16 to 17 patients per month, with considerable variation from month to month and year to year. The winter months, except December, usually showed an increase in calls, the summer months a decrease. The largest number was 49, recorded in January 1917, 46 in February 1916 and 42 in January 1923; the lowest was 4, in June 1921 and August 1924. 1923 was a bad year, with large patient loads in winter and summer, as was 1917; 1924 was the lightest year, averaging only 7 days and 10 patients per month, indicating that perhaps Protar was permitting himself to slow down. This trend continued for the first few months of 1925, before his death. In the period covered by the ledger (March 1915 to March 1925) he saw and treated more than 2000 patients. Many received more than one medication at each visit.

Although Protar gave up farming on a large scale, he lived on the fruits of the land. His diaries attest to the importance of his garden, as he reports the harvests with pleasure or dismay. Though his crops varied somewhat over the years, he usually planted cabbage, beans, peas, all kinds of root vegetables including carrots and beets, onions, cucumbers and tomatoes. He mentions corn, but in quantities suggesting that it was used for feed, as were sunflowers, "pea-straw" and hay. He had apple and plum trees, and green grapes were said to grow around his house, from which he made juice. He also made apple cider vinegar and sauerkraut, pickled his cucumbers, salted down meat and put eggs in water-glass. He kept chickens, setting about a dozen eggs every April, and occasionally he mentions goose eggs. In 1903 he obtained some goats. He had three horses: a young mare named Tissa, an older mare, and Harry, his faithful servant. At Harry's death in 1913, he wrote one of the longest of all the diary entries and one of the few expressing his personal feelings:

"Harry, born April 8, 1885, died Nov. 11, 1913, served from May 4, 1893."

Protar's life on the island was deliberately simple and austere. He gave up shaving and haircuts and let his beard and hair grow long. At various times he considered vegetarianism and abstinence from smoking (cigars). He had, of course, no indoor plumbing. Woodstoves were used for cooking and heating; the one in the main room of the house rested on a rock where one back foot was missing. He had to concern himself with storm doors and windows, summer kitchen (back stoop) and winter bed, pump water or snow water, and stovepipes on and off. He used candles for light, a kerosene lamp for reading. He apparently bathed at times in a large barrel on his front porch, but the season for this was short, with "first bath" at times not reported until August. He washed his own bedlinen, but his personal clothing was laundered and ironed by a neighbor, Mrs. Boyle, who also provided him with bread and an occasional meal. In his German-American cookbooks he 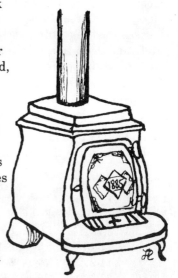 marked dumplings of all kinds the most often, and also salads, chicken, fish, pork chops, calves liver and honeybread. When he was feeling under the weather, especially toward the end of his life, he did not care for solid food, but would take a kind of corn-starch pudding prepared for him by Mrs. Bonner or her sister, Mrs. Boyle. Friends and neighbors helped with the haying, fencing and firewood, especially his closest neighbor, Patrick Bonner, who also attended to some of Protar's household chores.

Besides his medicating and gardening, Protar concerned himself with another feature of island life: the weather. After 1895 he kept daily, detailed notations, which constitute the bulk of his diaries. He noted the type of weather, temperature ranges and wind direction. Although he used some subjective terms: "ideal fall day" "very raw" "deadly heavy air," the entries were quite precise; some were annotated later with red or blue pencil and summarized, as if he were reporting them. However, it is understood that the island Catholic priests undertook the job of weather observation, and one, Father Zugelder (1899-1905) even set up equipment for a weather station. Most likely it was the Parrot heritage that induced Protar to undertake these studies. He developed a system for grading severity of storms based on the comparative and superlative Latin forms of the word "furious," in later diaries abbreviating them to F, FF or FFF. He remarked extremes such as snow depth, sudden temperature drops, early or late frosts or thunderstorms, droughts, excessive wetness, and smog (from forest fires), and such related events as grasshopper plagues, lightning strikes and fires, mail by boat or ice, and road conditions. Occasionally he added descriptive running headings, continuous from month to month, such as from November 1919 to April 1920: "Inordinately — persistent — sharp — long — win — ter!" Sample entries for six months, illustrating the four seasons, are given below.

Note: Protar's own ( )s are given as ( )
Author's asides in the diary notes in [ ]

December 1900

1-7	Mostly calm, cloudy 32-35
8	F storm, SW, then snow flurries 30-20°
9-10	Storm FF NW 10°
11-13	Storm F 10-32°
14-22	Weak E wind. 32-48°
23	Storm SE Rain, fog 38°
27-28	Storm and snow SE to W 30-22°
29	Storm! FFF SW 30°
30/31	Storm F NW 10-16°
5	Made sauerkraut. Cabbage frozen.
7	Eggs!!! From young hens.
10	Storm door installed!
14	Storm windows installed!
26	Set up juniper.

April 1912

1-6	Fair, W to SW 10-16°, 36-55°
7	Storm, light snow SW to N 32°
8-12	Fair N 28-36°/13 Storm E 28-36°
14	Storm, steady drizzle rain E 38°
15-20	Fair, Raw, very raw! N to SE 36-52°
21	Storm-rain-thunderstorm E 46-38°
22	N 30-34°/23-25 Fair SW 54°
26	Steady rain storm-thunderstorm S 50°
27	Storm N 26-38°/28-30 Fair N to S 36-54°
1	Started eggs.
5	Pump open, just 3 months [since]. Last ice mail.
6	Salted down meat.

April 1912/continued

11	Storm windows S off.
13	Patty Roe†
19	First mail boat (Gasoline launch).
21	First thunderstorm.
27, 30	BI Lumber Company first [in red].
29	Storm door S off.

June 1907

3	11 am to 4 am uninterrupted rain
10	Smog
11, 15	Rain
16/17, 26	Storm
17, 18	FFF thunderstorm, cloudburst, W 80-73°
27-30	WSW 48-57° Raw

10	Planted dwarf fir [?]
27	Hens (Orpington) set 10 eggs
28, 29	Seeded upper field — Early's
1-16	Off and on evening heat
17-30	Daily evening stove

July 1907

7/8, 19	Thunderstorms, rain
15	Big steady downpour
25/26	FF storm NNW, rain (not above 77° all month)

	GRASSHOPPER PLAGUE
19	4 chicks
27	Paul arrived harbor
29	Paul visit here

August 1907

1	Steady downpour
6, 15/16, 23	FF rainstorms, heavy air
11, 14/15	Thunderstorms, rain S-SW 77°
21/22	SW 31°
26	Foggy, rain SE 73°

	GRASSHOPPER PLAGUE
21/22	Frost. Cucumbers! Tomatoes!

October 1903

2-4	Thunderstorm. F rainstorm NE + SE 62-78-64°
6/7	FF Thunder-rainstorm NE
7	Drizzle, steady rain
9/10	Calm, moon — bright 26°!
10-14	Ideal fall! Calm 63-32°
15	3 pm to 11 pm 16th = steady rain 60-38 E-SW-W
22, 23	FF storm W-NW-N 48-20°
23-27	Storm, mostly bright 58-26° W-NW
28-30	Mostly stormy 40-58° SW, bright, ideal autumn

2	Brought in corn
5	Melons — none ripe — brought in
10	Tomatoes brought in
12	Pulled onions and beans
19	Started roots, ended 24th.
29	Set hyacinth, tulips, narcissus.

The surviving volumes of Protar's library reveal much about him. Only two are wholly in English: the US Dispensatory (medicinal substances and how to use them) and the Gospel of Buddha. His German-English medical, veterinary and gardening references, cookbooks and dictionaries show hard use. Besides the stage plays purchased in the 1870s, the collection includes a number of German classics (some of which are dramas) by Goethe, Schiller, Lessing and Hebbel, as well as a German cultural history and volumes of philosophy and poetry. One volume by Maxim Gorky and seventeen by Leo Tolstoy, all in German translation, were apparently brought by Dr. Bernhardi on his annual visits. Protar subscribed to German-American newspapers, but only the annual almanac supplements survive as his diaries. It is not known how extensively he corresponded but his concern about mail delivery in the winter, as recorded in the diaries, suggest that receiving mail was important to him. Friends and former associates from Rock Island came to Beaver Island during the summer, some originally to visit Protar but later to fish and vacation. Dr. Bernhardi, alone or with his family, came almost every year, as did Paul Kersch, who for a number of years after Protar's death continued to come to the old house, which he had bought along with much of the property.

Protar withdrew not only from urban life when he came to Beaver Island, but he withdrew from all social life and became a virtual hermit. With few exceptions, people interested him only if they needed his help. He had no use for strangers and curiosity-seekers, and did not tolerate intrusion into his private world. He could be and often was gruff and forbidding. This was not, as has sometimes been assumed, because he feared detection by some external political force, but because the focus of his existence, of his whole being, was on the inner man. All his efforts were devoted to the battle raging within him. He was by all accounts, including his own, a troubled, struggling, unfinished man.

Why did a well-educated, cultured, successful man-of-the-world choose to lead a precarious existence in virtual exile on a remote island? According to his notes going back to his Dresden days, this was the logical culmination of a life-long struggle with spiritual and moral values, a striving for simplicity, purity and selflessness. In his twenties he had written "Shall I be my brother's keeper? Assuredly, yes!" and "Give much, take little." In his eighties he marked in a philosophical work a passage attributed to St. Paul:

It is not I who live, but God who lives in me. This is attained only by one who overcomes his own person, one so deeply internalized that he finds his greatest happiness not in self-satisfaction but in sacrifice, in giving without the need to take in return, in godlike spontaneity.

These principles could not be satisfied in the public world, in the theater or in journalism with their sham, vanity, ambition and use of power, which he abhorred. Renouncing all this, he affirmatively sought a simple, virtuous life, earning and sharing his bread with others who worked hard for theirs, which he felt he found, "1893 to ??" on Beaver Island. He deliberately sought a world unlike the urban, cultured, literate one he had known before, one to which he could belong and contribute but where he could also find the solitude he needed to come to terms with his inner self.

It has been said that Protar was a follower of Tolstoy in his social and personal philosophy. However, it appears that he had developed his principles earlier, independently, out of his own experience, and then found them affirmed in Tolstoy's writing. Protar is less radical than Tolstoy — he did not need to expiate a profligate youth based on inherited wealth — and differs from Tolstoy in another major respect: his philosophy is applied only to himself, it is entirely internal, he has no desire to share, publicize, or impart it to others, even though all his work and competence had been in the area of communication and persuasion. He did not want to reform society, only himself, which he saw as an all-consuming task. And unlike Tolstoy, he was ultimately able to live his principles, to disencumber himself entirely, and to spend the final third of his life in this attainment.

Protar continued this caring, frugal, solitary life until the very end. The 1925 diary ends on February 28 with a "furious NW snowstorm." On March 2 he dispensed his last doses of cough medicine and acetanilid. He had marked, and most likely reread, a passage for March 3 (the bookmark was here) in his devotional book: "The only preparation for death is a virtuous life." The next morning when Patrick Bonner came to build up the fire, Protar was snoring heavily with his eyes open and could not be roused. He slipped away quietly before Pat could return with Dr. Palmer from St. James. When Dr. Bernhardi had left Protar the previous summer, they realized that they would not see each other again. "When I die, you will know?" Protar had asked. "I will, I will," Dr. Bernhardi had replied. He died at 7:45 pm the same day, March 3, 1925 in a Chicago hospital.

Protar had asked that his body be slipped into the water between Beaver and High Islands without any kind of ceremony. "No coffin, no black dress, no grave digging, no

flowers are necessary. One strong bag and a heavy stone will cover all requirements ... and four friends — in summer to row, in winter to drive me on the ice out, and to let me slip into the water, is all what is necessary." However, since state law did not permit this, he was buried on his own property, at the edge of the forest, near his home, and the islanders took up a collection to erect his tomb.

Protar's home, as well as his tomb, remain as memorials to him and his chosen way of life. Recently entered in the National Register as a historic site, the home stands across the road from the Centennial Farm of his good neighbors, the Bonners. It includes the house, part of the barn, the outhouse and chicken coop, gates and fences, and some of the trees he planted, among them apple, plum and cedar. The house, valiantly maintained by the Beaver Island Historical Society, is intact, except for a new roof and shutters to protect the windows, the loss of its back stoop (Protar's summer kitchen), and much of its siding, revealing the massive log construction. It is a serene place where one can see and feel, hear and smell the ambience that drew Protar to this rough, lonely but peaceful land. A shady avenue of old trees and rail fences leads westward to his tomb.

An essential part of Protar's creed was to be a humble, secret giver. But his gift shines through the gruffness and reticence with which he armed himself against the world and illuminates his existence. Though a conscientious man who had the courage to live his principles to the exclusion of the external world, he would most vehemently have insisted that he was too imperfect to be called a saint, as he would not tolerate being called "doctor," or being exalted in any way. The islanders, who could never know him well, accepted him for what he was to them, but the quality of their memory, in simple gratitude and admiration as a friend, shows that they did understand and value the man he was trying to be.

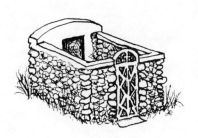

THE RISE AND FALL OF THE FISHING INDUSTRY

by William Cashman

The history of Beaver Island's fishing industry is the history of the fluctuation of the island's population. Fishing settled and unsettled the island. The first white inhabitants were fishermen, and one legend claims the early Ojibways settled here because of the excellent fishing. When the fishing was at its peak between the 1880's and the 1930's, the population reached its post-Mormon high of nearly two thousand. At this time St. James was renowned as the country's leading fresh-water fishing port and its harbor was ringed with net sheds, reels, ice houses, fish buyers, docks, and fish boats. But when the fishing gave out most of the people moved away.

Commercial fishing in upper Lake Michigan didn't begin until the 1830's. Before that time there was neither sufficient population close by to constitute a sufficient market nor quick, regular transportation to the large cities. In the 1830's three factors coincided to give an impetus to the industry: the fur trade disappeared, forcing a large group of businessmen to develop a new source of income (the American Fur Company, for example, began an extensive whitefish operation); the population of America and of Michigan began skyrocketing; the movement of these people past Beaver Island produced a comparable growth in Lake transportation, making the major markets more accessible.

This impetus would have led elsewhere if the fish hadn't been there. But they were: at this time upper Lake Michigan was the most fertile fishing ground in the world. The fishermen probing it from Mackinac Island found it almost unbelievable. These fishermen were mostly immigrants, and the amazement they conveyed in their letters home made these waters fascinating to the fishermen of Europe, and gave them the same appeal, particularly to the Irish, as Sutter's Mill had to the settlers beyond the frontier.

In the 1830's Mackinac Island was the only center of civilization in these waters, and the arriving fishermen made it their base. They ranged out for the fish and returned with their catch. The grounds extended past the Beavers. As the fishermen in their small sailboats ranged farther their returns to their base became less frequent. Soon they were building shanties on the islands in the Beaver Archipelago to serve them during the entire season.

Some of them clustered together around the excellent harbor on Beaver Island, primarily near the cooperage and store Alva Cable built at Whiskey Point in 1838. This fishing village grew in the 1840's and would have continued a term of regular growth if James Jesse Strang hadn't interjected his domineering presence towards the end of the decade.

At first Strang and his Mormon followers lived in peaceful coexistence with the wildly independent fishermen, but as their power and numbers grew they began to take steps to gain exclusive control of the island. By 1852 Strang had driven the non-Mormons off, and during the rest of his reign he effectively kept them from using even the most remote of Beaver's shores. If they accidentally drifted ashore he quickly devised a suitable punishment, sometimes, as was the case with the two Martin brothers, stripping them of their supplies and setting them adrift.

The Mormons had come to farm, but they couldn't ignore the fertile waters; probably between fifty and a hundred of them made fishing their profession. Strang taxed each fisherman ten dollars. He initiated this rule before he'd gained control, and the Bennetts' refusal to pay was behind his quarrel with them. Strang also organized the collection and burial of fish refuse, improving both the sanitation and the soil.

During Strang's reign fishermen continued to pour into upper Lake Michigan's fishing grounds, but they were forced to camp on the other islands. They were anxious to locate on Beaver because of its fine harbor. Finally Strang was shot and his followers run off by a mob of fishermen and traders from the Mackinac Island area. While this mob was provoked by land speculators eager to acquire parcels on Beaver, it was the fishing grounds which gave value to the land they were after.

As soon as the Mormons were gone the fishermen on the out-islands flocked to Beaver. John Bonner, for example, who was camping on Gull Island, was roused by an Indian, who told him, "Big man shot! Big man shot!" Within the hour Bonner set out for Beaver.

The word went back to Ireland that here was a place, quite a bit like the Emerald Isle, where land was plentiful and the fish even more so. In the next decade a fishing camp sprang up every mile or two around almost the entire island, particularly on the east and west shores. The pattern of life which emerged from these settlements included fishing as the central ingredient. Those who fished full-time in the summer frequently spent the winter cutting wood because, until 1877, the first boats in the spring would buy cordwood from them, providing them with the capital to start fishing again. Some of the early families had two houses, one on the beach for summer and another in the cedars they retreated to for winter.

Even those who came to farm, such as Vesty McDonough, began fishing as well as soon as they arrived. The fish were too plentiful to be ignored. All the early families who lent their names to the points and bays of the coast, such as McCauley, McFadden, Kelly, Kilty, Left, and Greene, were families of fishermen.

In the 1840's and 50's these fishermen used gill nets exclusively. They were made by hand during the winter, frequently at netting bees. When the Parish Hall was built much later, netting bees were held in it. The twine was knotted securely around a "mesh block." Until 1874 stones were used as weights, and strips of wood were used as floats. The floats and stones were tied on as it was being set, and were taken off when it was pulled out.

These nets, which are like underwater fences, were set within sight of shore because it was necessary to locate and tend them with a small Mackinac boat of from 18 to 26 feet. When the wind died they were propelled by a "white ash breeze." In this early period these nets were tended every four or five days, so that the fish were drowned.

A gill net catches fish when they run into it or try to swim through it, tangling their gills and fins. The size of fish caught is determined by the size of mesh, unless it's set slack, in which case it will catch smaller fish as well.

The Allers boys stringing nets.

The gill net season started in the spring. The nets were pulled out around the 10th-15th of June when the fish went into shallow water. For a month pound netters had an exclusive shot at the fish. Then the gill nets were set again in the dark of the moon early in July. The biggest lifts were made during two weeks late in the fall when the fish return to their spawning ground. These lifts went on every day until the ice was made. It was the hardest season to work, but the most productive.

Early gill nets were about 200 feet long and contained a pound and a half of twine. Thirty nets were considered a good rig. A 4½ inch mesh was used for whitefish, 5 or 6 inch for trout. The older the nets were, the more they shrank and the smaller the mesh became. They were set along ridges a few fathoms from the crest. A man who knew the bottom could take twice as many fish as a novice. In the 1880's over 50 boats from St. James set these nets, usually near or on the bottom in up to 400 feet of water, although when the habits of the fish changed they might be set near or at the surface.

Pound net pile drivers.

In 1859 Captain Waggley revolutionized the industry by bringing a pound net to Beaver Island. It was set in 18 feet of water in St. James Harbor, and was not a success, but the next year it was set off High Island and the catch was remarkable. This Scotch invention is extremely effective. It's constructed of heavy cotton netting, and has four parts: the lead, the heart, the tunnel, and the pot. Except for the tunnel the webbing extends from near the bottom to just above the surface, held erect by long poles *pounded* into the lake bottom.

The pounding of the stakes requires a pile-driver (a 20 foot tripod with a pulley on the top and a channel for a weight to rise and fall) on a raft. A crew was needed to hoist the weight — unless Big Owen Gallagher or Barney McCafferty were handy. Then the stakes were pulled out again to protect them from the ice, using hand winches on barrel rafts. Occasionally the stakes were abandoned, and there are stories of them turning up at the wrong moment, punching through the bottom of a boat.

A pound net could be set in water up to a hundred feet deep. Stakes for this depth were made by splicing ordinary stakes together. The lead was placed near the shoreline and the pot in deeper water. These nets were usually set singly, but they could be set in strings, the lead of one attached to the back of another. To lift them a boat gets inside the open pot and hauls in the twine from all sides. When the fish are close to the surface they're scooped up with a dip net and placed in barrels or boxes, or simply shovelled into the bottom of the boat. Two to three tons was a good daily lift.

In 1881 a pound net was set in deep water and was so successful that within a few years they were all set in water at least forty feet deep. After the spring and summer fishing they were cut down, and sometimes spliced together, and set for shoal water fishing on the spawning ground.

Within a few years of its introduction, the pound net could be seen ringing Beaver Island and here and there on Trout, Gull, High, Hat, and Garden, and in the shallow water between Garden and Hog. Between Beaver Harbor and Kelly's Point there were a hundred pound nets in the 1860's. It at least doubled productivity, and productivity was the key concept of the time, radiating to the Beavers from the state and the country. Under the spell of this concept lake whitefish, which was one of the highest quality food fishes ever discovered in the world, was exploited with the same unreflective energy as Michigan's great forest of white pine or the native copper discovered in the Upper Peninsula's Precambrian Rocks. To equate these fish with their fate, one could think of them as Beaver Island Buffalo.

The pound net's productivity made some observers worry. Not only was it effective in taking fish, but because the mesh size was unregulated and frequently small (it averaged three inches but was fished as small as an inch and a quarter), masses of small fish were frequently captured.

Some old-timers accuse the pound net fishermen of occasionally "cleaning the grounds." The charge was that they'd move down the beach and dump their catch behind a swale instead of dipping out the fish they wanted; then they'd pick out the few salable fish (sometimes only one in twenty) and leave the rest. When the grounds were cleaned there'd be no fish and no pound nets set for two or three years, but then the fish would come back. It's claimed this happened three times between 1894 and 1930.

Another type of net that was moderately used each spring was the seine. Seines were operated in St. James Harbor as late as the 1920's by the Sendenbergs and by the Larsens, who used it for perch. Before 1875 it had been used for herring and whitefish, and a few times each decade for suckers. The seine is strung between two winches which sit 150 feet back on the beach and a few hundred feet apart, so that the net loops out into the water. Like a gill net it requires weights and floats to hold it in an upright position. The winches are cranked simultaneously by a man or two on each. The first pull might bring only 25-30 pounds of fish, but it "sets the table" by stirring up the bottom. Fish from throughout the harbor could be seen following the net in. The next pull might catch 200-300 pounds, and later pulls have produced 1500-1800 pounds of perch.

Seine netting ended when the harbor was dredged. Previously it had been open to gill nets, which had kept it clear of weeds. But Captain Allers and C. G. Chief Vandenberg unofficially banned gill nets in the harbor. Hook and line fishing was so good the next few years (around 1931) that people came to the island from all over each spring to catch their year's supply of perch. They were so plentiful that many fishermen baited two hooks on a line; that way they could catch 60 pounds of 1¼-1½ pound perch in an hour. But when the harbor was dredged the sludge was allowed to float back in; it was like tilling and fertilizing the bottom, and a bounty of weeds sprang up between the fish and their food.

Before that time gill nets had been used in the harbor both before and after the ice was made. Before the ice a man might net as much as 1500 pounds of perch in a day. After the ice a long cedar pole attached to a line was used to string the net. It was given a good push from an initial hole. When it stopped a small hole was cut above it and it was given another push. Sometimes the ice stopped the boats a week or two before the mail plane was scheduled to start its run; in the 1920's this was the only plane, so when Bud Hammond arrived he found as much as seven tons of perch waiting to be flown to Charlevoix.

Chub nets were never used out of St. James. Captain John McCann gave this small-mesh net a trial one year but too many young trout were trapped in it so he hung it in the rafters of his shed and let it rot.

Beaver Island fishermen caught some herring, perch, suckers, sturgeon, and lake trout, but from 1830 to 1890 whitefish was king, completely dominating the catch. Towards the turn of the century lake trout became more plentiful than the dwindling whitefish, and despite trout themselves dwindling they remained more abundant into the 1940's.

Some sturgeon always were taken, although most of this presently highly valued fish was destroyed before the 1880's because there was no market for it. Fishermen hated these large (up to 400 pounds and 9 feet in the early days) and scaley fish because they wrecked their nets and were suspected of eating spawn. When they were found in a net they were usually gaffed out, injuring them sufficiently so that they'd soon die. It's estimated that 10,000,000 pounds minimum of sturgeon were destroyed in Lake Michigan. Now this fish sells for a dollar a pound.

In the 1880's a few Indians began marketing the fresh sturgeon they took with 25-30 foot long, detachable-tipped spears. They would paddle about in the smooth water near the out-islands watching for a sturgeon lying motionless on the bottom. In 1885, 450 lbs. a day (7 fish) was not an uncommon catch for a man with a spear.

After the turn of the century the total catch remained relatively stable, but three factors place this apparent stability in perspective. First, the most valuable fish began a consistent decline before the turn of the century (whitefish before 1880 and lake trout in 1892), so that a progressively larger portion of the catch was made up of inferior fish. Second, the population continued to increase, so that the availability of fish to the consumer declined (this wasn't true of land crops). And thirdly, technical improvements were made in all equipment so that the area fished and the extent it was fished increased rapidly without an increase in the product.

The productivity of the fishing grounds around Beaver Island followed slightly behind the cycle that held for the Great Lakes in general:

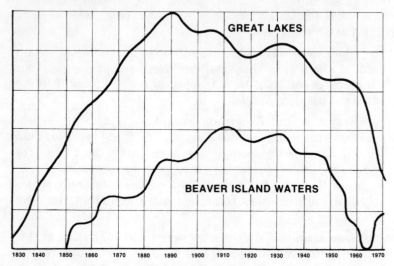

Although this decline in prime fish started in the 1890's, Beaver Island's waters were so abundant it wasn't noticed for 40 years. The fishermen's attention was focused on a constant stream of improvements instead.

After the introduction of the pound net the first major change was made by the introduction of the steam tug. Previously all fishing was done from sailboats or rowboats. Steam tugs allowed the men to work in any weather and more than doubled the number of accessible grounds.

The first steam tug purchased by a Beaver Islander, shortly after the appearance of the "Kitty O'Niel" had started them dreaming, was the "Clara A Elliot," a 52 foot tug purchased by the Martin Brothers in 1882. It used a crew of six to ten men and could set five and a half miles of gill nets. It was such a success that other tugs began appearing, culminating, in 1894, in James McCann's "Margaret McCann," the queen of the fleet.

James McCann was one of the first to come to Beaver Island after the Mormon exodus. A cooper by trade, he avoided fishing until 1865. By the 1880's he was using several boats and arranging for the sale of his and others' fish. During the 1880's his catch averaged slightly under half a million pounds of fish a year.

But when the "Margaret McCann" arrived he decided it would be all he needed. Under the charge of his son John, using up to a 15 man crew, this 69 foot beamed boat could set ten miles of gill nets. For ten years it alone averaged 250 tons of fish, taking 395 tons in its best year.

A steam tug required a considerable investment, and only a few of the fishermen could afford one. Frank Left bought three in succession, the "Lily Chambers," whose engine and boiler were purchased for $50 by James Martin for his second steam tug, the "Shamrock," the "Little Matey," and the "A & M Link," built in Charlevoix by Swales. The Sendenbergs, with the help of Frank Cornstalk from Garden Island, built themselves the "Two Sisters." Pooler built a dock in front of the McDonough store for his steamer, the "Sydonia," and Jimmy Gordon first had the "Badger" at Sendenberg's dock and then the "Lily May" at Pooler's. Many of these steamers were pressed into tug duty in Buffalo during W.W.I. The Martins, who'd had the first, also had the last, the "Evelyn M," built in 1922, which had a 5 ton boiler. It was converted to crude oil in the early 40's.

When the tugs were operating they were fueled by coal. Their owners bought coal either in 50 ton amounts from a freighter from Cheboygan, the "E. J. Leway," or in smaller amounts from Mike McCann, John's brother. When the township acquired the Beaver Island Lumber Company dock the pilings were covered and the fishermen who used it had as much as 100 tons of coal dumped on it, at $4 a ton. During the winter they frequently took a sled down to the coal pile for a few hundred pounds to heat their houses. The biggest boat to come to Beaver Harbor was the "Paisley," which dumped coal on Mike McCann's dock.

In 1896 the McCanns installed a power lifter on the "Margaret McCann," which enabled them to set and pull more nets. It was the first one used out of St. James, and on a still day its large open gears could be heard for two miles.

Gill nets had been improved considerably by the turn of the century. First, leads replaced stones for weights, and corks replaced the carved, notched, and oiled floats. A Canadian named Deering came to the island to fish and had the first leads and corks. He lost his equipment in Johnny

Martin's Dirty Hole. The Dirty Hole is a gamblers' specialty. About 15 fathoms deep a mile off Point LaPar, nets set in it might catch a considerable number of whitefish, but they might just as well come up with a load of stumps and branches, or not come up at all.

A few of these new nets washed up and the Martins found them and began using the leads and corks. By 1885 half the gill nets had them. At about the same time twine companies began offering prefabricated netting, so that the number of nets a man could set no longer depended on how many he could make.

These improvements, especially the power lifter, increased the average number of nets a small boat could set to about 12 boxes. Each net was 500-600 feet long and five to six feet in breadth. Nets were stacked four to the box. Twelve boxes could be set in an hour and a quarter. Some boats set as many as 95 boxes on a spring day.

A crew using 12 boxes had their work cut out for them. Each day they had to go out to their ground, lift 12 boxes worth of nets over the bow, take out and sort the fish, and set 12 boxes over the stern; of the 12 boxes set, half were those just pulled out of the water, but six boxes of cleaned and dried nets were set also. Then the six boxes pulled but not reset would be brought ashore.

Net reels.

These nets were made of cotton fibre, which would rot if kept in the water too long. So they had to be cleaned and dried. The easiest way to clean a net was to drag it in the water behind the boat, one man feeding it out and another taking it in so that 50-100 feet of it was in the water. Every month and a half to two months they were boiled in water containing soap and tan bark — tamarack — for a half hour. Then they were dried, sometimes by being hung in a shed but usually by being wound on a net reel.

As many as eight nets could be wound on a reel. As they were strung, the leads and floats were removed and placed in the netbox. When the net was dry it was ready to go back in the water again.

Captain Allers, Captain Roen, Gus Mielke, and Irwin Belfy on the "Schnoden."

Art Larsen started another revolution when he had the "Estonia" built by the Van Saus, a family of boat builders from Garden Island, in 1922: it was the first diesel tug. The Sendenbergs followed with the "Sylvia S," then the "Agnes S," and finally the "Bobby Bill." After the "Ocean" Andy Gallagher bought a boat Ulysses McCann built, the "McCann Jr.," and then had the "Elizabeth G" made from cedar crooks from Squaw Island. Ulysses then built the "Venus II" for his brother Roland. Harold McCann built the "Liberty" for himself. Mike Cull, after co-owning the "C & C" with Hugh Connaghan, first had the "Betty C" and then the "LaFond." Raymond and Lester Connaghan bought the "Seagull" to replace the "Nautilus," built by Gus Mielke. Mielke also built the "Silver Moon," which Barney Martin bought after owning the "Mary M." Irwin Belfy called his boat the "Silver Star"; Wilfred O'Brien's was the "Gypsy"; Mal Gordon's the

Meadow Mac"; Ernie LaNeau's the "St. Teresa"; Charlie Martin's the "Globe." On Garden Island Matt and Pete Jensen had the "Panther" while on High there was Nanigaw's "Sea Gull," Little Joe's "Margaret," Alec Cornstalk's "Stella," Jack Anthony's "Sarah," Napon's "Elizabeth," and Paul Kenwabakise's "Whitefish."

In the early thirties the "Elizabeth G," "Evelyn M," "Estonia," "Agnes S," "Betty C," and "LaFond" found the fall fishing so good down by the Foxes that the freighter "Rambler," also built by Mielke, was pressed into service to take their catch to the Booth Dock in Charlevoix and return before dark with supplies. The grounds were too good to leave. While the run held, the "Rambler" became a floating social center each night as the fishermen celebrated their good fortune.

The finished "Schnoden."

The gas engine was the most important 20th century improvement, allowing each fisherman to have his own powered boat, but there were many others. Nylon nets were stronger, lighter, and free from rot. The pound net was revamped by closing the top of its pot so that it could be set in any depth of water, held in place by anchors and floats. Floats were now made of hollow aluminum; during the war, people walked the beaches of the other islands each Sunday looking for them. Later plastic floats replaced the metal ones.

The introduction of the "submarine" or deep water trap net in the late 20's resulted in such tremendous catches of whitefish that a law was passed banning it from depths greater than 80 feet. Subsequently this was reduced to 50 feet.

These modernizations simplified the work of the fisherman, but they increased the investment required. And the life remained hard.

From the start, only the hardy could take it. The early fishermen had to rise before dawn and put to sea. Their traditions were based on fighting the Atlantic, so they bravely faced Lake Michigan's storms. They were known for pulling their nets in weather in which everyone else huddled in their homes. Needless to say, they didn't always return.

Even on a calm day lifting nets was backbreaking work. Dressed in oilers, sometimes splashed by a freezing spray, the men were exhausted by hauling ton after ton into the boat. After lifting the nets they had to clean them. If anyone had a free moment there were plenty of other things for him to do.

On shore the nets had to be reeled, the other equipment maintained, and the fish either cleaned and salted or taken fresh to a fish buyer. For each crew of three or four there was a shore hand to help, usually a young boy who dreamed of going to sea or an old man glad to preserve some contact with the work he loved.

Before the turn of the century, when most of the equipment had to be made by hand, there were over 500 fishermen, and each of them was good at making nets, boats, and barrels. But as newer equipment was developed the jobs became diversified into independent specializations. After the turn of the century the number of fishermen operating out of St. James dropped to below 200, but the business still supported three quarters of the island's population.

This support partially came from the subsidiary industries, some of which date back to the beginning of fishing here, such as coopering. Through the 1870's the entire catch was salted and shipped in half-barrels, which held about a hundred pounds. Most fishermen could make barrels themselves, and some were quite good at it, but they could never make enough, so a cooper was needed. Beaver Island had its first cooper in 1838, and always had one — and usually several; there were twelve at Cable's Bay in 1850, for example — into the twentieth century.

The early fishermen could sell their fish to either a trading schooner or a trading post. The schooners visited the out islands at odd intervals, buying salted fish and selling such supplies as twine and pilot bread. When the schooners reached Chicago the salt fish were sold and resold and eventually rinsed for a day or more to remove the salt.

The schooners competed with each other and with the posts. One schooner sought to impress the fishermen by paying in silver dollars, or, as was more usually the case, with a silver dollar. When James McCann opened his store in 1885 he also had a schooner to sail between it and the scattered camps.

In the late 1880's steam tugs enabled fishermen to bring their catch back to St. James each day so that the fish no longer had to be salted. Reed and Murdoch continued buying salt fish, but other Chicago buyers encouraged the fishermen to use ice to ship their fish fresh. Ice houses quickly sprang up around the harbor, and soon some of them were the biggest buildings on the island.

Taking ice in the winter for storage was done here as late as 1958, using the same tools and techniques developed at the turn of the century. A slide, a plow, a big saw, tongs, and a horse were needed. After the ice on St. James harbor was cleaned a channel was cut, starting at the slide and leading outwards. Each new chunk that was cut was pulled along the channel and up the slide, and either directly into the ice house or onto a waiting sleigh. After it was stacked in an ice shed ice was chiseled off the top to fill the seams, and it was covered with sawdust. To be used it was crushed in a one-lung gas powered crusher.

By 1954 Beaver Island was the only community in the United States still to be putting up ice. At the end (1958), only three tiers were put up since nearly everyone had a refrigerator. Instead of the usual crew of 24-30 men, James Gallagher, "Young James," who directed the operation, had only Jimmy Floyd, Paul — "Young Paul" — Kenwabikise and his son John, and Brian, Dale, and Alex Cornstalk to help him.

Fish boats at Whiskey Point.

As ice replaced salt fish first the fish car and then the fish box replaced the half barrel. The fish car was wooden, galvanized inside, and had two or four wheels. It was packed with 500-1000 pounds of fish and ice and pushed aboard a freighter, sometimes with the aid of a block and tackle. It had a shelf midway so that two fishermen could each send their catch in it, or two kinds of fish from one man could be sent.

The fish car was excessively cumbersome and soon yielded to the fish box. The box was better than the barrel because it could be mass produced more easily, and, using nails instead of hoops, cost far less. It was so inexpensive, in fact, that it was classified as a nonreturnable.

The manufacture of fish boxes supported small mills on Beaver and Garden — over 10,000 were needed each year. The boxes were made of balsam, spruce, or cedar boards, some of which were cut from trees boomed to St. James from the other islands. The middle side planks were made longer to provide a handle.

Another subsidiary business was the production of fish oil. By boiling fish refuse a highly refined oil is produced for lubricating small motors. In the 1880's 50 to 100 barrels of it were produced each year. The last men to operate this business were Sam and Frank Floyd in the 1920's.

By this time Beaver Island had a reputation for such productivity that some of the largest fish buyers from New York and Chicago acquired dock space in the harbor. The Fulton Fish Market competed with the Booth Fisheries for the prime spot, but over the years Booth came out on top. Booth thus set the price paid for Beaver Island fish.

Booth started on McCann's dock, which now belongs to the Beaver Island Boat Company. When Booth moved to the Gallagher dock, which was in front of the Post Office, Mike McCann became an agent for the Standard Fish Company. The Lakeside Fish Company operated on the dock that was in front of the Marine Museum, but then replaced the Walker Fish Co. on Beadle's dock, which was where the yacht dock is now. The Beadle Fish Co. operated here well before 1900.

When the C. C. Robbins Fish Co. put a buyer on Art Larsen's dock Booth had set a price of 11¢/pound for trout. In order to get into business, Michaels, the buyer, offered 18¢, but there were few takers. Most of the fishermen had signed contracts with Booth in exchange for an advance which they needed to start the season.

The fish buyers, by operating as bankers, had the fishermen at a disadvantage, and generally they received an even lower percentage of the retail price than the farmer. But sometimes these forced contracts worked in the fisherman's favor. One year the Floyds agreed to sell to Booth at 5¢. The fishermen selling elsewhere for much more laughed at them, but then the bottom fell out of the market and they couldn't get much more than a penny a pound. The Floyds, though, were able to hold Booth to its contract.

Irwin Belfy's "Silver Star."

Occasionally the McCann's would load up the "Idler" with fresh fish and take them directly to Chicago, but the usual alternative to selling to a fish buyer was for the fisherman to ship his fish on the railroad. Fish buyers without local agents provided the fishermen with tags. When a fisherman was ready he'd call the various markets for their up-to-the-minute prices, select the appropriate tag, and put his tagged box on the ferry to Charlevoix. Art Larsen, for example, shipped a lot of fish to both Dierssen and Johnson and to H. Grund & Sons. In Charlevoix a truck picked up the boxes and took them to the railroad. On the train they were re-iced. The peddlers that bought from these Chicago markets developed favorites and would ask for fish shipped by fishermen they believed packed their boxes tighter. These fishermen got their checks back first.

When fishing was just reaching its peak on Beaver Island observers downstate had already been predicting the demise of the industry for over fifty years. For example, the first report of the State Commissioner and Superintendent of State Fisheries maintained in 1874 that

" . . . These lakes in former years, and even now after years of improvidence and waste, produce millions of whitefish annually. Yet the catch is very appreciably diminishing."

Ten years later, in the fifth biennial report, the indictment had grown more serious:

"The fishing grounds are one after the other fished out, and then new places sought where the same process is repeated. If each ground, as it becomes unprofitable for large operations, was actually abandoned and allowed to rest, it would undoubtedly be slowly restored to productiveness by natural processes, but this seldom happens."

A note of panic crept into the tenth biennial report (1892):

"The cupidity of the selfish fisherman should give away to his judgement, if he reflects and understands that a few years more of present modes of fishing must leave the waters of the great lakes nothing but mere waterways for the passage of our lake commerce, their valuable fisheries having passed into that stage of decay which now distinguishes Lake Ontario."

Because of this apprehension a replanting program was begun in the late 1870's to bolster the lake trout supply. When the Charlevoix Hatchery was opened around the turn of the century, with a 30-40 million fish egg capacity, all the Beaver Island fishing boats took on a new member of the crew: the federally hired spawn taker. During the spawning season he collected spawn, at first being paid $5 a day but then being paid by the quart. Andy Gallagher, "Andy Mary-Ellen," who took spawn on the "Margaret McCann," took almost 700 quarts in one day during the brief season. Dan "Turner" Boyle worked on James Martin's tug and would fill up a pork barrel each day.

The spawn taking process started with the capture of a mature male trout. After stroking the spawn out of a female the spawn taker would milk the male in the same manner, stirring the milk into the spawn with the fish's tail. After as much spawn as the hatchery could handle was gathered, the more conscientious takers continued gathering and fertilizing spawn, dumping it back on the natural ground.

The fertilized spawn was received on Beadle's dock and sent across to the hatchery. The fry were raised to fingerling size and then released on the same grounds they'd come from. These grounds were marked by "honeycomb," rocks of the St. Ignace Formation in which imbedded salt crystals had dissolved, leaving the hard stone riddled with holes in which spawn was sheltered from its predators: suckers, lawyers, and menominees. The prevalence of this honeycomb rock is one of the reasons these grounds were so fertile. A similar shelter for spawn was provided by the clinkers dropped by passing freighters.

In the early part of the century the commercial fishermen performed an experiment in Charlevoix under the supervision of the federal government, and discovered that in natural spawning only 0.5% of the eggs hatched out. But if the spawn-takers were careful the yield in the hatchery was 98%. This meant the restocking program was 200 times as effective as nature in producing young trout.

The spawn taking operation enabled the fishermen and the fish to remain in equilibrium. To maximize production the fishermen wanted to see it increased. They continually asked for more hatcheries, but in the mid thirties funds ran out and the entire program was suspended. Suspecting they'd fallen victim to the more lucrative sport fishing industry, the commercial fishermen predicted the end of their occupation.

Efforts to replace the program with a series of rigid regulations failed. The rules were neither economically feasible, co-ordinated with other states, nor geared to the fluctuating habits of the fish.

The decline was gradual at first. To an observer it could best be measured indirectly. Children of fishermen no longer followed their fathers into the profession, breaking a chain that had held for three and four generations. Lost or damaged equipment seemed pointless to replace. When Art Larsen's nets were destroyed in the Armistice Storm of 1940, for example, he sold the "Estonia," which had been blown on the beach, and went sailing. On High Island the storm hit so hard that Paul Kenwabikise, the last of a group of fishermen which had included George Nanegah, the Naponts, and the Batistes, who'd also built many fishing boats, was forced to leave; he lost his nets, his boat, the "Whitefish," the wreck of which washed up on Garden Island, and all his buildings as waves of near tidal proportion swept the beach. Paul moved to Beaver to work for the Gallagher Brothers, Willy John and Ed Bowery, who were among the last pound net fishermen. The days when all islanders were given free live red lake trout each Friday were over.

The killing blow to the industry was delivered by the lamphrey eel, which arrived in Beaver's waters in 1936. The lamphrey is a living fossil which dates back 450,000,000 years. It's the only living representative of the "jawless

A Palm Sunday ice shove moved the Booth shed off the dock.

fishes." Under its predation the lake trout were exterminated and the annual production of whitefish in Lake Michigan dropped from the millions to 25,000 pounds. The former foods of these fish grew in number, completely upsetting the ecological balance. On Beaver Island the traditional ceremony of blessing the ships had to be dropped; there were no more ships. All the net sheds ringing the harbor stood idle. The fishermen either died, moved away (the population dipped below 200), or sat dreaming of the island's lost glory.

Attempts to control the lamphrey began in the mid 40's. In the 50's barriers to lamphrey spawning streams were erected, but they proved ineffectual and were abandoned by 1958. Since 1953 10,000 chemicals were tested for their ability to destroy lamphrey larvae in their bed. One of them, TFM, seemed promising. In 1960 TFM was used in 109 lamphrey spawning streams, and by 1963 Lake Michigan's lamphrey concentration dropped to 20%, and to 10% by 1964. In 1965 restocking the lake began.

The restocking program has been designed to benefit the sport fishermen exclusively. Little interest in a rebirth of the commercial fishing industry has been shown. But on Beaver Island, whose economy has adjusted to a new base, three families have begun fishing. Don Cole and Paul Kenwabikise have been using trap nets around Beaver, Garden, and High. In 1976 Ernie Martin returned to the business of his father, grandfather, and great grandfather. They find fishing profitable, but they offer only a little hope to those waiting to hear once again thirty proud tugs leaving St. James each dawn.

87

NOTES ON ISLAND LOGGING

by David Gladish

Cordwood

Ninety-nine years ago, in 1877, the wood-burning Lake steamers began switching to coal. It did not happen over night. These transitions never do. But it brought changes. For Beaver Island the advent of coal meant the phasing out of a lucrative sideline of the fishing community — a sideline that had flourished since steamers began plying the Lakes.

This industry, of course, was cordwood. We have an early record of it in the notes James Strang made when he visited Beaver Island, in 1849. In the Gospel Herald of May 17 of that year, after mentioning the cedar and other softwoods on Beaver Island, Strang says, concerning firewood,

> Wood sells at $1.50 per cord to the boats which stop at this port on their way up and down the lakes, and the wood business makes employment for a good number of hands now, and it will increase yearly. Wood choppers get fifty cents per cord, cash, for chopping wood, and the timber is about the right size to make it a good business. Any smart boy, of a dozen or thirteen years of age, can chop a cord of steamboat wood in a day, and do it easy. They cut the wood a little short of four feet, and do not split it very fine.

So the trade in cordwood must have been well established on Beaver Island before the middle of the 19th Century. Steamers took on wood at both ends of the island, indeed — at the Harbor, and at Cable's Bay.

It takes the right things at the right time for any industry to flourish. The coincidence that made the Beavers an important source of fuel was that the islands lie approximately midway between Detroit and Chicago by water, are right on the shipping lanes and have the fine natural harbors where the fuel was loaded.

And there had to be the basic resource — wood. Continuously throughout its entire history, Beaver Island has been forested with around fifty square miles of timber, much of it hardwood. As other Michigan wood reserves dwindled in the latter 19th century, the out-of-the-way Beavers remained lush and unexploited by commercial logging. Wood was in abundance.

Another commercial use of cordwood from the island, after the days of wood-burning steamboats, was "coke wood." This was wood to be burned for charcoal for the foundry fires of the Antrim Iron Works, and it was exported between 1920 and 1945. It was hardwood "bolts," or logs in the round, except that everything over four inches in diameter had to be split. Local men were paid to cut and work up this coke wood, at the rate of about $2.25 a cord, piled, Archie LaFreniere recalls. A standard cord, incidentally, is a stack eight feet long by four feet high by four feet deep, or 128 cubic feet of logs. The $2.25 represented what today must seem an astonishing amount of work, especially without chain saws. But at that, it was more than double what the choppers got per cord in the 19th Century!

The remains of a heap of this wood, which Patrick La-Freniere says was cut in 1929/30, and which never got shipped, can still be seen at French Bay.

To bring the subject of cordwood up to date, it continues to be a minor source of income on the island today, and one which shows signs of growing with the dwindling of oil and the boom in nostalgia for woodstoves and fireplaces. Though the real market for cordwood today is strictly local, one experimentally-minded islander took a pickup truck of birch firewood "across" recently, drove it downstate and sold the wood at a shopping center, more than paying his expenses.

Tan Bark and Ties

If any industry exploited Michigan's resources more than logging, and sooner, that industry was the fur trade. Tanning furs requires tannin, and a great source of tannin is hemlock

Loading tan bark at Iron Ore Bay.

bark, so here, too, the woods of Beaver Island responded to an off-island need. From the first, there have been fine stands of hemlock on the islands, and so "tan bark" has also had a place among exported forest products.

To quote Laurence Malloy, an oldtimer who had a keen interest in island history and was a life-long resident, here is an illuminating glimpse of the tan bark operation.

> The tan bark . . . was a great commodity in those days for the tanneries . . . [It] was loaded out here [at Camp Four], and elsewhere around the island, wherever there was a swamp adjacent to the beach. They felled the trees, let the logs lay there and took the tan bark off the trees, in the early days, and . . . sold the tan bark.

Swamps, both adjacent to the beach and inland, are typical of island topography, thanks to various causes: residual wetlands left by prehistoric lake levels, a hard sub-stratum of glacial clay, and the industrious damming operations of the beaver. This condition has meant that there are very few sections of land without some swamp, and consequently cedar has always thrived. This is the northern white cedar, or arbor vitae.

For a period in the late 19th Century, when logging operations on the mainland were using narrow-gauge railroads to haul their logs to water, cedar was much in demand for railroad cross-ties. Many cedar ties were cut and shipped from Beaver Island for this purpose. During the months when the Lake was frozen, ties made a good off-season occupation for the fishermen, who lived all around the coasts of the island. They would go back into the woods, hand-hew the cedar ties and carry them out to the beach.

The ties (and also hemlock for tan bark) were stacked in growing piles, till spring. Then, according to Mr. Malloy, schooners would pick them up right from the beach "wherever they saw a pile of logs." Thousands of ties, he says, were made on the island, before the Beaver Island Lumber Company came. The remains of one unfortunate man's heap of ties, once "a quarter of a mile long," according to Patrick LaFreniere, are still turning up near Kilty's Point, rejected by the buyer because they were not of uniform length.

However, the market for cedar ties passed, probably before 1890. While cedar was excellent for resisting rot, it was too soft and light to hold up as well as denser woods treated with creosote.

Saw Mills Before 1900

The mainstream of Michigan's great 19th-century logging boom practically passed the islands by. Pine was what the industry was after. Competition was fierce, and a thirty-mile journey by boat would have come right off the top of profits from island pine in those days.

Yet there were some mills on Beaver Island, operating at various times in the 19th Century. Cable, of Cable's Bay was selling cordwood before the Mormon invasion, and some say he operated a mill there, too.

It is also likely that the Mormons operated a saw mill during their brief occupancy at mid-century. They used some sawn lumber and shingles even for their log buildings, and it certainly would have been to their advantage if they made it locally. In any case, four of them gave their occupation as "lumberman" in the United States Census of 1850.

After the Mormons were gone, Waggley ran a mill at Cable's Bay, as recalled by some old-timers.

Land records show that in 1873 two partners named Boardman and Sweet were buying land on Beaver Island, according to Mrs. Helen Collar. Boardman had been operating a mill on Grand Traverse Bay in 1847, so they were no doubt buying the island land for timber. It was probably in the late 1870's that Sweet's Mill was put into operation, at the southern end of Beaver Island's Sand Bay, in the clearing at the end of Hannigan Road. The mill continued in operation at least until 1891, because in that year Patrick Maloney was killed in an accident there.

Timber was brought to Sweet's mill on wooden tracks that ran on what is now Hannigan's Road but used to be called the Old Mill Road. The logs were not hauled by engine on the wooden railroad, but probably by horses and oxen.

In 1898, Charles Tilley, a carpenter by trade, and his partner, Noe Stebbins, brought another mill to Beaver Island. The mill stood not far from the Harbor at St. James, near Lake Drive, and it supplied the lumber for the King Strang Hotel, among other structures. This mill was later sold to Wilbur Gill, according to Sybil Larsen, Charles Tilley's daughter.

The Beaver Island Lumber Company

By the turn of the century, few stands of virgin pine were left in Michigan. At the same time, a new phase of logging had begun. Hardwood was in growing demand, for furniture and flooring.

The size and quality of wood on the islands had never been a problem. Stumps thirty inches across can still be found — relics of the last century — and squared logs in old log houses may have a face of twenty-two inches. Patrick La-Freniere, who has been almost everything in his life, including a timber cruiser, says that the typical 40 acres of timber on the island in the old days could be expected to yield 440,000 board feet of hardwood.

The Beaver Island Lumber Company's Mill.

At last the island forest, still practically virginal, began to look profitable from a mainland point of view. The Beaver Island Lumber Company installed a large mill, built offices, shops, lodgings and docks, laid a railroad and employed 125 men, according to the BIOGRAPHICAL HISTORY OF NORTHERN MICHIGAN (1905, p. 616). The company, the HISTORY says, "was organized on December 27, 1902, with a

One of the BIL Co.'s Engines with a load of logs.

capital of seventy five thousand dollars, W. E. Stephens being elected president, John S. Stephens, vice-president, and G. Kitsinger, secretary and treasurer."

The mill of Stebbins and Tilley was bought from Wilbur Gill, but not for the Lumber Company's use. It was where they were setting up their own mill, Sybil Larsen says.

By 1903 work was under way. The mill had "a daily capacity of thirty thousand feet of hardwood lumber" and 75,000 shingles. The first season they made 500,000 feet of lumber and over 2,000,000 shingles. Much of this went into their own numerous buildings on the island, the HISTORY says. The focal point was the mill and the main dock, on the Harbor, at the present site of the Beaver Haven Marina.

Many company residences stood along Freesoil Avenue (the Stephens brothers' home town was Freesoil). They are said to have rented for $3 a month. Fire and decay have taken their toll, but some of these company houses are still in use.

The company Boarding House, where L. J. Malloy said a man paid a painful $3.50 a week, is at present Dr. Sorenson's house, at the Harbor end of Lake Drive. McDonough's Store was the Company Store. The house back of the present Post Office, near Gatliff's, was the so-called "sleeping shack" for the Company. Clustered around the mill were big stables, a blacksmith shop and the railroad roundhouse, Malloy recalled.

You can see why the Company needed to provide a certain amount of inexpensive housing when you consider the wages. They seem appalling today and were probably not outstanding three-quarters of a century ago, for that matter. The monthly payroll for the 125 men was $3,000, according to the BIOGRAPHICAL HISTORY. This figures out to exactly $24 per man per month. L. J. Malloy remembers that "the lumberjacks worked for practically nothing." The majority of the mill crew got "between a dollar ten, dollar fifteen, dollar and a quarter a day, and that was for ten hours!" Patrick LaFreniere puts the wages for lumberjacks in those days at $16 a month, and less. And there must have been a gruelling pace to meet at times. One man in the shingle operation kept up by having his wife and young daughter stack bundles of shingles, Sybil Larsen says, so they were ready for him to bind.

In addition to the lumber mill and shingle mill, the operation also included a stave mill and a lath mill.

The markets for the lumber, shingles, staves and lath

Part of the Beaver Island Lumber Company's Work Crew.

were principally Chicago and Milwaukee. Schooners and steamboats were the carriers that took the products there. They moored at several company docks, where tram cars moving on rails brought the bundled wood.

The big dock, at the mill itself, stretched far into the Harbor. Not far to the east, where Clyde Fogg's house is now, there is another dock from which a great deal of cedar was shipped. A third, east of that, now used as the DNR dock, L. J. Malloy said, was built as a slabwood dock. Slabs for firewood were also shipped to Chicago and Milwaukee. Beyond the DNR dock during construction of a ways for vessel repair, they recently unearthed the rails of what could once have been a tramway serving these docks.

But the rails that brought the timber in from the woods started at the mill, angled across the present Beaver Haven property, turned by a huge burnpile behind "Judy" Palmer's house, and headed west, roughly along the alley that comes out at the site of the present Episcopal Church on Gallagher Avenue and the Back Road. Here there was a "Y," according to Mrs. Collar. Then the narrow gauge followed the Back Road to the Kenwabakise's home, where it curved towards the Donegal Bay Road. It went west to Donegal Bay along this road and then turned south. It ran behind Barney's Lake on what is now the Kuebler Trail, past Protar's Tomb, and past the west side of the airport. On the south side of the airport it met what is now the West Side Road, most of which was built on the railroad bed, forked where Hannigan's Road meets the West Side Road, and continued to Camp Five at Miller's Marsh. The fork ran about a mile to Camp Three.

The train brought logs — mostly hardwood — into town from four camps. Camp One was just south of the present

Township Airport, near where the West Side Road joins Paid een Og's Road. Camp two was a mile northwest of Fox Lake. Camp Three was as far south of the same lake. Camp Five was just south of Miller's Marsh.

Camp Four was just north of Iron Ore Bay, joined to Camp Five only by a wheelroad and not served by rail. Instead, the logs from Camp Four were moved to the lake and "boomed" to St. James — towed behind the tug *Ryan* within a floating enclosure made of logs fastened end-to-end in a circle. The towing operation took about two days and a half, Malloy said.

The reason this could be done was that most of the timber from Camp Four — called "the softwood camp" — was, as the name implies, softwood. It floats higher in the water than the denser hardwoods and lends itself to this type of booming. Hardwood was towed on the water, too, but each log had to be individually fastened by "dogs," hooks you drive into the log.

Rail transportation for the hardwood camps was, of course, more reliable. The engine could operate summer and winter, independent of wind and weather, with a snow plow in front when required.

The company went through three railroad engines in the course of their island operation. The first, called the "Nancy," came from a narrow-gauge railroad in the Benton Harbor area. It was old and required repairs sometimes, L. J. Malloy recalls. They acquired a second engine, called the "Pop." A third name appears in the Charlevoix *Courier* for October 26, 1966: "The Beaver Island Limited" — presumably the third

A fatal train wreck.

The Spirit of Beaver Island.

engine that the Lumber Company operated. It was in this engine that engineer David Chase was killed, when it was derailed on April 17, 1908, according to a note by Malloy on a clipping from the *Courier*.

Not that derailments were unusual. Eight to ten cars in a train were normal. Mrs. Larsen says that on their winding path through the woods the heavily-loaded cars would tend to spring the tracks on curves with their sideward thrust. If this went unnoticed, the train was in for trouble next time through.

Most trips, however, were successful, and besides hauling logs the trains also provided a handy form of transportation for islanders who wanted to hitch a ride south to pick berries and come back to town again afterwards.

At the harbor the engines were turned around on the roundhouse turntable for the next trip. There were "Y"s at Fox Lake and Camp Five. When the engine drove out one leg of the "Y" they threw the switches behind it so it could back up and out the other leg, the way you back and fill to turn an automobile around in a small parking place.

The train picked up the logs at locations called "banking grounds," where logs were piled, or "decked" on either side of the tracks. Camp Five, at Miller's Marsh, was the largest camp, and the banking ground would have logs heaped "fifty, sixty foot high," according to Malloy, and stretching a quarter of a mile, and "millions of feet of logs have been banked here by the Beaver Island Lumber Company."

Camp Number Three.

In the winter, logs could be brought to the banking grounds by sleigh — not only by the Company's camp, but also by private "jobbers" who logged small remote parcels not worth the Company's trouble. But much use was also made of "big wheels." The term applies to a pair of gigantic wheels, twelve feet high, with an eight-inch-wide steel tire. Logs were slung under the axle, and the rig could be pulled over rough terrain by horses or by oxen.

High Wheels.

Camp Five finished the Lumber Company's task. John Gallagher says that they came to harvest a certain amount of timber, and when they finished, they left. It was in 1915, according to the memory of "Judy" Palmer, whose family is one of the older on the island, that the Beaver Island Lumber Company took up their track to ship it back to the mainland. The mill was also dismantled and removed.

Other Logging in the 20th Century

There was still fine timber on the islands after the Beaver Island Lumber Company left. In fact, the last stand of virgin timber, called Big Owen's Woods, was not cut until around 1930. Quite a number of mills have operated, so that logging has been nearly continuous right up to the present.

Schweitzer's mill was in existence during the Lumber Company era, at the opposite end of the island, on McCauley Bay. It was bought by Garrett Cole, who operated it from 1912 until about 1926, making lumber and shingles. It was the hub of the community of Nomad.

In 1915, Mrs. Larsen says, Charles Tilley was operating his second mill, this time on the Back Beach, near where Freesoil Avenue meets Pine Street. He made fish boxes. Tilley sold out to Gus Mielke, who set up at the site of the Beaver Island Lumber Company mill and dock, on the Harbor.

Charles Pischner, who is a nephew to Mielke, and who worked for the mill in various capacities, recalls that it employed four men downstairs, three upstairs, and a man outside loading bolts on the conveyor, or "bull chain." Among the products of the Mielke mill were boxes for the South Haven vineyards, fish boxes which were shipped to Charlevoix, Petoskey and Cheboygan, birch boards to be shipped to Muskegon and turned there for dowels, and, of course, lumber. Mielke built the *Rambler* to ship his products on. Much of the timber for the box wood was cut on Garden Island, winters, and boomed to Beaver Island in the spring.

When Gus Mielke died, the family sold the mill to Elston Pischner. It continued in operation until the 1960's, and then it was sold to the Beaver Haven Marina.

But, going back to the teens, another of the mills operating then was on Sloptown Road, not far beyond the place where you see the microwave tower today. Dan and Hugh Boyle operated this mill.

In the late 1920's and the 1930's Michigan Maple Block Company was logging, though they had no mill on the island. L. J. Keller, their foreman, appears in the daybook from Nels LaFreniere's store as early as 1926.

According to L. J. Malloy, the company is supposed to have taken out over 7,000,000 board feet of logs, which were taken to Petoskey by the steamer *Stuart* at the rate of a trip every 24 hours. Their dock, angling southeast along a reef in Iron Ore Bay, was built by John W. Green in 1928. The cribbing of it can still be seen under the water. The logs were drawn onto the dock in tram cars by horses shod with rubber, Patrick LaFreniere recalls, to reduce damage to the planks and give better traction. It was usually wet or icy underfoot.

In a small operation on Darkeytown Road, Peter Johnson made "corks," or cedar floats for fishing nets, which he sold both on Beaver Island and the North Shore.

In the 1920's, William Ricksger operated a mill just south of McCauley Road, opposite Rose Connaghan's present house. The mill was powered by a Model T engine. It was where the Southern house stands.

There were two mills near Fox Lake, Spade's mill at the Fox Lake "Y", and Chancy's mill about a quarter of a mile in from the road.

"Shoemaker's" mill was on the east side of the island. According to Patrick LaFreniere, the boiler that can be seen on the beach about a half a mile south of Point La Par belonged to that mill. There was a banking ground over the beach near the north end of what is now the Wicklow Beach subdivision.

There was also a portable mill operated by Ed Young at the south end of Cable's Bay.

Tony Wojan operated three different mills in the 1930's and 1940's. The first was near the dock now belonging to the Beaver Island Boat Company. In the 1940's he set up a mill on the south side of Green's Lake, the foundation of which can still be seen. After WW II, Wojan and his sons moved this mill to the location in town that is now occupied by the Harbor View Motel. The boiler was the one that now lies at the north end of the motel, and it had formerly been the boiler for Cole's mill, at Nomad.

In about 1942, a man named Carpenter set up a mill in the clearing south of where the Lumber Company's Camp Three had been. Carpenter had first come to the island as millwright for Cole's mill, according to Don Cole.

Carpenter moved his mill to a location a half-mile north of Miller's Marsh to make maple rounds for croquet equipment when Ziebarth organized that operation, in the 1940's and 1950's. The mill also made tool handles, L. J. Malloy recalled, and lumber. It continued to be operated until the late 1960's by a long line of operators including Laurence McDonough, and ending with Art Brown. It has supplied squared logs and lumber for many a cabin on the island. The collapsed remains of it can still be seen, just beside the West Side Road.

Logging on Other Islands

Virtually all the islands have contributed to the logging industry in one way or another. Various mills have boomed timber from Garden and High Islands. Hog and Squaw have also been logged.

The House of David operated a large mill on High Island during their occupancy there, between 1913 and 1930. And a Texan named Northcutt operated a mill on Garden Island, at Northcutt Bay.

Logging at Present

In the 1960's, Bill Welke installed a small saw mill next to his landing strip to saw lumber for local construction, on demand.

In 1975, Walter Wojan and sons installed a mill, on the Back Road in St. James. During the winter of 1975/6 the Wojan crew cut an impressive supply of timber, which is banked at the mill and is being sawed for use in local building. Export is also a possibility, according to Walt Wojan.

There is virtually no virgin timber left on the islands, needless to say. But second-growth pine, hemlock, cedar, poplar, birch, maple, beech and oak are reaching an acceptable size. With the selective cutting methods used today, the stands should steadily improve.

This much-needed industry seems here to stay. Knock on wood!

THE RELIGIONS OF BEAVER ISLAND

by Terry Hart

Indian burial mounds, found at Beaver Harbor and built by ancient Indian tribes, indicate very early habitation on Beaver Island by man. Little is known of the tribe's religious beliefs, other than the conclusion that the mounds were built for burial and worship.[1] Much later, other Indian tribes settled on the Beavers, bringing with them varied and primitive beliefs.

Christianity may have come to Beaver Island as early as the 1600's. It is thought that a white settlement might have been started shortly after Champlain arrived in Canada in 1603. There are geological indications of extensive fields which were thoroughly cleared and cultivated at that time.[2] It is highly possible that if such a colony existed, Christianity was part of the culture, but there is no proof of this.

The first written account of the introduction of Christianity to Beaver Island is given by Reverend Frederic Baraga, a Roman Catholic priest, who was pastor of the parish in Arbre Croche, now Harbor Springs. After Easter in 1832, Father Baraga set out on a journey to Beaver Island to announce the Word of God to the Indians living there. He tells of that trip in a letter to the Leopoldine Foundation in Vienna, a group which provided support for his missionary work:

"I set out for that place with confidence in the Lord, Who has promised that He would always be with His servants until the end of the world.

"My heart beat fast as we approached the island. — I have a white flag with a red cross in the center, which I let wave when I sail to a mission so as to make the canoe of the missionary distinguishable. — As we approached Beaver Island we had a pleasant mild wind. The peaceable flag with the cross waved gracefully over my canoe and announced the arrival of the servant of the Crucified. When the Indians, who have marvelously sharp eyesight, noticed and recognized my flag from afar, the chief of the island had his flag immediately hoisted to the peak of his hut. My Indians, who were taking me there, immediately noticed the flag of the chief. Now I was at my ease, because from this I judged the good disposition of these islanders. As we came closer we saw many Indians hastening to the shore; nearly all the inhabitants of the island assembled there to welcome us; and the men gave two salutes with their rifles to show how very pleased they were by the arrival of the missionary on their island. Scarcely had I stepped on land when all the men approached me and in a friendly manner shook my hand, and then they led me to their village, (that is, eight pitiable huts of tree bark). I went first into the hut of the chief where many of these poor Indians, who could not satisfy themselves with gazing at a priest, had assembled. They have heard much about priests through tradition from their ancestors, but they have never seen one, and never has one come to their island.

"When one has to do with pagan Indians of this country, he must observe a certain ceremonial. I did not say forthwith what I had to say to them; we spoke about various other subjects, and at the end I requested the chief to call together a great council in the morning, (as they call it) where I want to consult with them about important matters. Accordingly they assembled in the hut of the chief and I made a speech to them in which I pointed out to them the necessity and usefulness of the Christian religion; and at the end of the speech I asked the chief for a reply; and he answered me very formally, through his speaker, that they are most pleased and consider themselves very fortunate in seeing a priest on their island, and that they wish nothing more ardently than to accept the Christian religion. — One can imagine the sincere joy with which this answer filled the heart of the missionary. I remained with them for some time and instructed them, and on May 11, (O boundlessly happy day) I baptized twenty-two of these Indians."

After a short visit to what is now known as Manistee, Father Baraga returned to Beaver Island:

"Here also they brought me all the idolatrous offerings, and I burned them. Here also they decided to erect a small church by the time I come again. For this time they built me a kind of chapel out of tree branches, sail cloth and blankets, in which, with grateful feelings, I said Holy Mass daily and thrice daily held religious instructions. Six more pagans were converted to the Christian religion and were baptized. However, there are still many pagans on this island who refuse to be converted. One day these came to my tent; some among them were almost entirely naked; a poor blanket around their loins was their only clothing; and one of them began to speak and made me a very silly speech in a shrieking monotone, in which he declared to me, in the name of all those who came with him, that they do not want to accept the Christian religion, but want to live and die in the faith of their ancestors. I answered him; and I hope that in time also these pagans, or at least some of them will be converted to the Christian religion. — I promised also the inhabitants of Beaver Island to visit them often."[3]

The next summer, Father Baraga visited Beaver Island at least twice. He found his converts there constant in the Faith but much harassed by the other Indians, who had not embraced Christianity. These pagans would not allow the Christian Indians to build a church under any circumstances and even threatened to burn it down if they tried to. Father Baraga brought gifts to the pagans — such as striped shirting, small scissors, needles, thread and tobacco — hoping to win their confidence. Finally, the pagans agreed to allow a church to be built on a parcel of ground, which they picked out far from their village. Father Baraga said of that, "I am grateful to Divine Providence that the matter was finally settled; I would rather have a church far from the habitation of these obdurate pagans, as we can hold our services in a quiet place without disturbance."[4]

When Father Baraga first came to Arbre Croche, all of Michigan was part of the Diocese of Cincinnati. In 1833, the Diocese of Detroit was created and Father Frederic Reese was appointed its first Bishop. Father Baraga was then notified that he was being transferred to Grand River. Before permanently leaving Arbre Croche, he went on one more journey to all his missions.

"Returning home, I also landed on Beaver Island, where I found, however, but a few Indians, because the most of them had gone to Canada to receive their presents from the English government (their annual custom). The church I found not yet completed because the heathens of this Island are still hostile to the Christian religion and tear down what is built, and thus hinder the construction of the Church.

"The Christians have, for this reason, come to the conclusion to leave the Island and to settle in L'Arbre Croche. I approved of it, leaving word to the absent Christians, that I desire them to move to L'Arbre Croche. This was the condition of the mission on Beaver Island when I last visited it."[5]

In the 1840's, the first white settlers came to Beaver Island. The Rochester Northwest Company, headed by Colonel Fisk of Rochester, N.Y. established a fishing station on Beaver Harbor. About the same time, Alva Cable of Ohio opened a trading house at Whiskey Point.[6] Reverend Francis Xavier Pierz, pastor at Harbor Springs from around 1840 until 1852, served the needs of the Catholic Indians on the islands.[7]

From 1845 to 1848, the islands were surveyed by the United States Government. Also surveying Beaver Island at the time, for other purposes was James Strang, leader of a sect of the Mormon Church (Church of the Latter Day Saints). Upon the death of Joseph Smith, founder of the Mormon religion, James Strang and Brigham Young both claimed to be his successor. This led to a permanent break between the followers of Strang at Voree, Wisconsin, and the main body of the Mormon church at Nauvoo, Illinois. In 1846, Strang claimed a message from God had described a "land amidst wide waters and covered with large timber, with a deep broad bay on one side of it."[8]

The following year, Strang and four of his followers visited Beaver Island and published full accounts of the island's resources and assets. In late July, four "chosen" families migrated to Beaver Island to establish a stake. By 1848, when the government put the islands on the land market, there were 20 or more Mormon families living on Beaver Island, increasing to approximately 150 the following year. They built homes, cleared fields and started to build their tabernacle, a 60 by 100 foot structure. Every member was required to contribute one-sixth of his labor towards its completion. On July 8, 1850, in that incomplete building, James Strang read, for the first time, a part of his *Book of the Law of the Lord,* which proclaimed the re-establishment of God's Kingdom on earth and designated James Strang to occupy the throne.[9] In an elaborate ceremony, Strang was crowned King.

The Strangite Religion had 4 sacred books: The *Bible,* the *Book of Mormon,* Joseph Smith's *Book of Doctrine and Covenants,* and the *Book of the Law of the Lord.* The last was said to be translated by Strang from 18 metallic plates, which he claimed to have miraculously discovered, and which he said were written in ancient times.[10] In an article furnished for the N.Y. Times by Charles J. Strang in 1882, we find this account of general domestic regulations:

"The discipline of the church in the matter of temperance and morals was very strict. The use of tea, coffee, and tobacco, as well as of liquors, was prohibited. The temperance laws of the State were strictly enforced with especially good effect among the fishermen and Indians. Polygamy was introduced during the winter and spring of 1849 . . . By-laws for the kingdom were adopted and published, and every household possessed a copy. They were very strict in all that regulated society, morals, and religious observances, and absolute obedience was enjoined. The seventh day was set apart as the Sabbath, and every person physically able was commanded to attend church on that day. The saints were required to pay one-tenth of all they raised, earned, or received into the public fund, and the tithing was used for improvements, taking care of the poor, and paying State, county and township taxes. No other tax was levied. Schools were organized and flourished finely. A printing office of sufficient capacity to print all the papers, books, pamphlets, tracts, etc., needed for the church was maintained, and became a strong arm in the association. No betting or gaming was permitted, but the rules were very liberal in the matter of amusements."[11]

All was not well with the kingdom, though. Troubles had been brewing between the "saints" and the "gentiles" since the Mormons first moved to the island, resulting in verbal and physical clashes. Then, on June 16, 1856, James Strang was shot. An account of the incident was given by Capt. Alexander St. Bernard, who was an officer on the United States steamer, *Michigan:*

"When we stopped as usual on one of our trips around the lakes, the complaints were so bitter that our captain made up his mind to arrest him (Strang) again, and he told me to find and bring him on board the ship. I went to the temple, first, where I was told that he had just gone home. I found him sitting in his room, with four of his wives, where he received me very cordially, and when I told him my errand, accompanied me willingly. He linked arms with me and we walked along talking pleasantly. Just as we stepped on the dock and started to walk down the narrow passage between the piles of wood, two of his enemies sprang from some hiding place and shot at him. He clung to my arm until they began to pound him with the butt end of their pistols, when he let go and fell, leaving me covered with blood from my head to my feet."[12]

After taking the murderers and some gentile families to Mackinac, the *Michigan* returned more than a week later with more of Strang's enemies, threatening to get him dead or alive; so Strang and a few of his followers boarded the "Louisville" and left for Voree, where Strang died from his wounds. The rest of his followers were driven off Beaver Island the following week.[13]

The Strangite Mormons were completely erased from Beaver Island, leaving behind only a few gentiles with fishing and lumbering interests there. They were soon joined by a group of Irish immigrants, looking for a new homeland. These new settlers at first took on jobs at the government lighthouse building projects or cutting wood for James Cable. Later, many turned to fishing and farming for their livelihood. The first few years were, without a doubt, difficult for Beaver Island's new residents, but they were good workers and had a strong faith in God to see them through, and He did. Many of them lived to ripe old age and their descendants still populate the island. It was said by one Gallagher, boasting of the islanders' longevity, that "they had to kill a man to get their graveyard started."[14]

In 1857, Father Seraphim Zorn, an Indian missionary, started erecting a log church on the southeast end of the island, two miles north of Cables Bay. This church would be called St. Ignatius.[15] That same year, the Upper Peninsula was made into a new Diocese in the Catholic Church and Father Baraga became its first Bishop. Although the new diocese did not include the Beavers, due to transportation difficulties in those days, the Bishop of Detroit asked Bishop Baraga to take over the administration of the islands, which he agreed to do.[16]

In May, 1859, Bishop Baraga again visited the islands he had originally brought Christianity to. He found there a settlement of Catholics, mainly Irish, but some German and French, eager to have their own church and priest. May 22, he said Holy Mass in a large schoolhouse and confirmed 24 persons. The Bishop then met with the men of the community to decide where and how a church might be built for them.

From Beaver Island, Bishop Baraga visited Garden Island, which was inhabited by Indians. All of them, it was said, were Catholics and were visited from time to time by Father Zorn, the same priest who pastored the Catholic Indians on Beaver Island. They had a chapel built of bark, but were on the point of building a new church of cedar at the time of the Bishop's visit.[17]

Bishop Baraga kept his promise to the immigrants on Beaver Island and appointed Reverend Patrick Murray as their priest. He also contracted Alexander Guilbeault in July of the same year, 1860, to build a church on the island. The agreement read as follows:

"Alexander Guilbeault agrees to build a frame church at Beaver Harbor on Beaver Island, Lake Michigan. The church shall be 50 foot long, 30 foot wide and 15 feet high inside, with one double door 8 feet high, and two windows in front, and 3 windows on each side. He agrees to make the door, but not the windows, except the frame of same. He also agrees to ceil the church inside, 3 foot from the floor, all around, except on the side where the altar shall stand. He further agrees to match and lay the floor, and to put a good shingle roof on the church with cornice all around, and to place and put clapboards on all four sides. He also agrees to put a steeple astride on the roof of the church, according to plan and instruction. Behind the church there shall be an addition of 10 foot, with a door from the church, and a door and two windows from outside, and one window in the loft of the church, according to plan and instruction. All the necessary materials will be furnished, he will have only the work to do; and for his work Bishop Baraga promises to pay him $250, as soon as the work shall be done. This work must be finished before the first of December next."[18]

One island legend says that a great hassle developed, as to where the church should be located, between the "fish-chokers," who wanted it in town, and the "hay-seeds," who wanted an inland site. A compromise was reached and the structure was built about a mile and a half from St. James Harbor.[19] The resulting church was named Holy Cross.

Father Murray built the priest's home at his own expense. Bishop Baraga referred to Father Murray as a "zealous missionary," who "has accomplished much good, principally in combating the vice of intemperance among his people." Father's temperance society boasted 90 members.[20]

By 1864, the Holy Cross parish had a membership of 79 families, and the first nun, Sister Dympheny, came to Beaver Island from Buffalo to keep house for the priest. She took cold and died the following year at Christmastime. Father Murray says in his death register that "she was from her youth an example of piety and a copy of virtue. May her soul rest in peace. She is buried at the foot of the high cross as befits a religious. She died at forty eight years of her age."[21]

Bishop Baraga continued to remember the churches at Beaver Island and Garden Island, and corresponded regularly with Father Murray. The Bishop purchased a 301 pound brass bell for the Holy Cross Church for $237.70, and instructed Father Murray to send the old iron bell to St. Ignatius. He had gone through much trouble to locate a teacher for the Indians of Garden Island, only to have the inhabitants, to his great dismay, refuse to accept the instructor. So he placed the teacher on Beaver Island instead.[22]

In the summer of 1864, Bishop Baraga embarked from St. Ignace on the schooner "Rutland," owned and sailed by John B. Bonner. He sailed to Beaver Island for what would be his last visit before his death four years later. This schooner was the same boat that carried the lumber from Traverse City to build the Holy Cross Church. A reproduction of a restored portrait of Bishop Baraga can still be seen in the Holy Cross Church in a frame made from oak taken from the "Rutland."[23]

Two years later, Father Murray was transferred to Alpena and replaced by Father Peter Gallagher. Father Gallagher was from Ireland and spoke fluent Gaelic. Island humor tells that one old woman is said to have stayed alive three years, just waiting to say her last confession to the Reverend Gallagher in the ancient tongue.[24]

The need for priests in those days was so acute that formal training was cut to the bare minimum. In only two years time, Gallagher was trained by one of the diocesan priests, ordained and stationed on the island. The Holy Cross Centennial Publication relates an interesting story about that:

"When Ignatius Mrak became Bishop of Marquette in 1868, he was well aware that his predecessors had ordained a few men who had very little formal training. Bishop Mrak was afraid that some of these men might not be fit for the work that they were doing, so he set about on a visitation of the Diocese, and where he found such a man, he gave him a test to determine his ability. Should such a priest fail the test he was retired. News of this reached Father Gallagher and being one of those ill trained priests, he was a little worried about the outcome of his examination. He finally prepared the people for his possible removal. All went well until the day that Bishop

Mrak arrived on the island for the purpose of testing Father Gallagher. The people welcomed him with open arms, but waylaid the boat captain and threatened him. They told him that if he did not leave the Island at once and take the Bishop with him, they would burn his boat. The poor captain was completely at loss as to what to do and so he ran after the Bishop and explained the situation to him. Bishop Mrak was so disgusted by the whole affair that he promptly left the Island. And upon arriving back at Marquette the first thing he did was to write off the Island from his administration and lay it into the lap of its rightful ordinary, the Bishop of Detroit."[25]

Father Gallagher pastored the Holy Cross Parish for 32 years, and then died of ptomaine poisoning. During his time on Beaver Island, he was also in charge of St. Ignatius on the southeast end of the island. He would go there every third Sunday to say Mass and stay overnight at the home of John Sullivan. Due to difficulties which arose, Holy Mass was discontinued at St. Ignatius and the building was converted into a school. The people continued to use the place for congregational praying when they couldn't attend Mass at Holy Cross.[26]

It was also at this time in the history of Beaver Island that the first recorded service of the Episcopal Church took place. On Tuesday, August 14, 1877, the Burial Service was read for Henry Clifton by the Reverend George Whitney, "an Episcopal Minister who delivered a practical and impressive sermon." It is said that Reverend Whitney was the first non-Roman Catholic Clergyman to have been on Beaver Island, and that he had come from Harbor Springs for the service.[27]

The Franciscan Fathers from Harbor Springs or Petoskey continued to look after the Catholics on High and Garden Islands, visiting them two or three times a year. Three of the Indian missionaries who attended the islands were Father Servace, Father Pius and Father Dorotheus Philipp. The High Island Church, *The Assumption of Our Lady,* was constructed in the 1890's.[28] Upon the death of Father Gallagher, care of Beaver Island also fell to the Franciscans, and Father Bruno Torka came over occasionally from Petoskey to serve the Island's Catholics.

On July 4, 1899, the Diocese of Grand Rapids, which had been created seven years before, sent Father Alexander Francis Zugelder to pastor Beaver Island's Holy Cross. Father Zugelder's six years on the Island were most fruitful. One of

111

the first things he did was to petition Bishop Richter for the Dominican Sisters of Marywood in Grand Rapids to come to the Island and teach in the schools. Therefore, that year, Sisters Clementine, Genevieve, Hildegard and Gabriel came to the Island and taught in the McKinley and Sunnyside grade schools. The first convent was next door to the parish hall, until a new one was completed next to the church in 1901. The rectory which Father Gallagher had occupied was in such poor condition that it was torn down and Father Zugelder lived in a house across the street from the parish hall. Father wanted a new rectory near the church, and the Holy Cross Centennial Publication tells how he accomplished that goal: "the story is told how the men were somewhat reluctant to get started on it. An old woodshed stood somewhere nearby and Father set it afire and then ran and rang the church bell giving alarm. The men came quickly only to see the old shed in ashes, so they all stayed for a day's work on the new house. And thus the new rectory was started. Father built a veritable fortress once he got started. The old house has walls twenty inches thick."

"Father Zugelder was also instrumental in getting the telephone cable to the Island. It was also during his time the pastor became a weather man. The tall tower was built in the front yard and from it by means of ropes and pulleys the priest raised storm warning flags by day and lanterns at night to warn the ships at sea."[29] He also had the church enlarged to twice its size.

Father Zugelder was a dynamic, energetic man. Many old island pictures show him with the islanders and Dr. Protar, who had arrived in 1902, at their picnics and other gatherings. One old Gaelic gentleman, who had trouble pronouncing Zugelder, called him Father "Shoeleather."

In 1905, he was transferred to Beal City, and the care of Beaver Island again fell into the hands of the Franciscans from Harbor Springs. According to the Franciscan History, *Heralds of the King,* "a Franciscan residence was established in the stone rectory, with two fathers of whom one was pastor and superior and the other attended the Indian missions on High Island and Garden Island. The first pastor was Father Pascal Foerster, 1905-1907, the second Father Norbert Wilhelm, 1907-1910."[30] Father Wilhelm had the present parish hall built, and planted the trees which stand in the church cemetery.

By 1910, the Island was flourishing with farming, fishing

and lumbering. The population reached a high of 1,095.[31] At this time, a diocesan priest was again sent to the Island. Father James Malone was an easy going Irishman, well loved by his parishioners.

That same year, the Beaver Island Lumber Company, (who it is said donated the materials to build the parish hall), erected the first Protestant church on Beaver Island just north of where the present Episcopal Church stands. It was built for their Protestant employees from Freesoil, Michigan. After the lumber company left, Dr. Ruth, a Methodist who was the first summer tourist to build a cottage on Beaver Island, and Mr. Ricks, held some summer services until Dr. Ruth's death in 1930. Dr. Ruth and Mr. Ricks also fenced the Protestant cemetery. One summer Gillis Larsen brought a Lutheran minister across the lake to baptize his granddaughter and the youngest Tilley girl. After 1930 the church was moved, and is now the Gatliff home.

In August of 1912, still another religious sect entered the history of the Beaver Islands, when a portion of High Island was sold to the House of David from Benton Harbor. This religious group was founded in 1903 by Benjamin Purnell, who claimed to be the "Seventh Messenger" referred to in the Book of Revelations in the *Holy Bible*. The members of the House of David believe that before the year 2000, the world will be shattered by a cataclysm from which they alone will emerge with bodies intact to live a thousand years of heavenly life on earth. Their men never shave or have their hair cut. All property was owned in common and their rules included vegetarianism and celibacy. High Island was purchased by the House of David for lumbering and farming.[32] They established a thriving community there, with two mills, a bakery, and many homes, including a large round house consisting of nine rooms. This group did not seem to have the difficulties that the Mormons did with their neighbors, and, on a whole, got along well with the Indians and Catholics on the Island.

Back on Beaver Island, upon the transfer of Father Malone, Father Edward Jewell was sent as a replacement. Father Jewell had been an Episcopal Minister in Manistee, and converted to the Catholic faith, after the death of his wife; he came to the Island the father of three children, two daughters and a son. At least one daughter lived with him and attended school on Beaver Island. An eloquent speaker, Father Jewell was also well-liked by his parishioners. He played with the children and sang songs with the old folks, and was a jack of all trades besides. One account tells how "Father Jewell carried the islanders through a severe epidemic of diphtheria when no doctor could come from the mainland, by telephoning constantly to Charlevoix for instructions, and using his own common sense and a natural skill in medicine, sharpened by experience."[33]

Around 1915, a few years before the ruins of what had been St. Ignatius were torn down, a new church was being built on Garden Island by Hugh Gallagher, Burleson Northcutt, Mr. & Mrs. Meds Jensen, Peter Nielson and the native Indians. The church was dedicated to the Sacred Heart. Only 11 years later the use of that church was discontinued.[34]

Father Leo McManmon succeeded Father Jewell in 1921. He was a young man and only child of his Irish parents, who therefore lived with him at the rectory. Father McManmon was very mechanial, building radios and having one of the first cars on the Island. The sisters also acquired their first automobile during his pastorate. He put electricity in the hall and in the Church to replace the old oil lamps which had provided the only light for 60 years.

After eight years of service, Father McManmon was replaced by Father Edward Neubecker, a young religious man. His cousin, Margaret Neubecker, came with him and kept house for him. During that time, The Brothers of the Christian Schools of the St. Louis Province received a gift of land on Beaver Island. After inspecting the property, it was decided to erect a building to serve as a summer home and for retreats. It was named Brother Domnan Lodge, in regard for their old teacher and friend, Brother Domnan.[35]

The Beaver Island population was slowly declining and by 1930 stood at 540 residents. On High Island, the House of David members had left, leaving only their caretaker behind. On Garden Island, the discontinued Catholic Church was sold to Lawrence Malloy, who in turn sold the property to Arthur Redderson for one steer. The United States had just fallen into the Great Depression, and its effects were felt on Beaver Island. It is said that many parishioners tithed the Church in cords of wood.

The Depression did not dampen the spirits of the islanders, who took the occasion in 1932 to celebrate the centenary of Bishop Baraga's first visit to Beaver Island. Father McLaughlin was pastor of Holy Cross at the time, and the festivities included a Pontifical Mass, with a week of ball games, horse races, Indian dances and a water carnival.[36]

Father McLaughlin came to Beaver Island in 1931, bringing along his brother, an excellent piano player. In keeping with the island tradition of nicknames, Father's brother was referred to as "Eddie, the Church."

When Father McLaughlin was transferred in 1935, his replacement was another man of Irish descent. Father Eugene Fox served the Island three years and was then replaced by Father Anthony Bourdow, who brought his mother with him. During Father's pastorate of two years, the interior walls and ceiling of the church were covered. He was also well remembered for his weekly movies at the hall. Father Francis Branigan, who followed him, was the last Diocesan priest to serve Beaver Island.

In 1942, at the invitation of Most Reverend Joseph Plagens, Bishop of Grand Rapids, the Order of Friars Minor Conventual took charge of the Holy Cross Parish. Father Fabian Keenan was sent as pastor. The use of the church on High Island was discontinued and the property was sold for $100. The bell, altar, pews, wooden candlesticks and statue of the Blessed Virgin were brought over to Beaver Island. In a letter written by Father Keenan in 1953, five years after he left Beaver Island, he remarked, "I was well pleased with what the people did for the church over the period of six years that I was up there."

He was succeeded in the summer of 1948 by Father Giles Berthiaume, O.F.M.C. After completing three years as pastor, Father Berthiaume died suddenly on September 25, 1951, as a result of a heart attack while on vacation in the Porcupine Mountains in the Upper Peninsula. The vacant pastorate fell to Father Joseph Herp, O.F.M.C.[37] Father Herp could be seen playing saxophone at the dances or umpiring ball games at Whiskey Point, all in his cassock. He taught the young people swimming and water skiing, but also watched over the spiritual life of his people closely.[38]

The Holy Cross Parish then took on a major project, as related in the Centennial booklet:

"In November 1957 work on moving the Church began. A moving concern from the mainland was engaged for the work and the owner of the *Mackinac Islander,* a freight boat plying these northern waters carried the heavy equipment over to the Island free of charge over operat-

115

ing expenses. The Church was cut in two just beyond the second window from the front and each section was moved separately. It was noticed that the foundation beams under the 97 year old section were pretty far gone and as that part was raised up on a dolly the old church really looked her age. But with fervent prayers and much luck it made the trip to town. During the period of moving, services were held in the parish hall."[39]

Except for the short-lived Protestant Church built by the Beaver Island Lumber Company in 1910, the Catholic Church was the only one on the Island. Over the years, however, the protestant population began to grow. In 1954, the Rt. Rev. Dudley B. McNeil, then Episcopal Bishop of the Diocese of Western Michigan, came to Beaver Island on vacation. A paper submitted for this history tells the following story:

"He no more than got off the boat when he was approached by a summer resident who asked if he would celebrate the Eucharist. A permanent resident, the mother of Mrs. Spaulding, had not made her Communion for years. Of course, Bishop McNeil was delighted to hold this service. After much hurrying and scurrying — including a flight to the Detroit area by David Wilson to get some Prayer Books — the service was held in the Spaulding home on July 11, 1954. This house is now the home of Dr. and Mrs. Joseph Christie. It was anticipated that perhaps six persons might attend, or a dozen at the most. A "Coffee Hour" had been prepared with this in mind. This service had the house filled to overflowing and the doughnuts were broken into small pieces so that everyone present might have at least a bite. This was the beginning of the summer mission; Bishop McNeil held services for the rest of the month of July. He was followed in August by Fr. Arthur R. Willis who, with his family, was vacationing on the Island.

"By July 25, 1954, St. James Day, a Chapel had been established. It was the old Tilley house on Freesoil, now owned by Kathryn and David Wilson, who being Episcopalians, offered it for this use. Through the concerted efforts of many resident Islanders and summer residents, the old house was refurbished and furnished as a chapel. The altar appointments were handcrafted from native materials. The Very Rev. Glen Blackburn, Fr. Willis and Carl Felix made the Cross, Altar and lectern. The pews were rough planks given by the Wojan Mill which were placed over new empty fish boxes given by Elston Pischner. At the end of the summer, these were replaced by cathedral chairs given by Warren Townsend. However, the fish boxes and planks served as auxiliary sitting as long as the "old farm house" was the Chapel.

"By the summer of 1961 there were movements afoot for the erection of a more permanent building. Further impetus was given when Mr. and Mrs. Warren Townsend, Sr. gave two lots on the east side of Gallagher at Oak Street to the mission. A fund raising campaign was launched under the auspices of Bishop Charles E. Bennison of the Diocese of Western Michigan. Bishop McNeil had resigned from this post in 1959. Dr. Paul Nelson was in direct charge of the campaign. Contributions were made from the Diocese and by summer residents, and very importantly, by Islanders — not only those who were not Roman Catholic but many Roman Catholics as well. During the summer of 1963 a Bellaire log chapel was constructed. On September 1, 1963, this Chapel was consecrated by Bishop Bennison."[40]

As the Protestant population on Beaver Island continued to grow, the need was felt for an interdenominational Protestant Church. The same year that the new Episcopal Chapel was erected, a church of that type was started. Mr. Phil Gregg tells the story:

"The first service of the infant Beaver Island Christian Church was held on Easter Sunday in 1963 in one of the units of Wojan's Motel #1. Attending this first service were Dr. and Mrs. Haynes, Mr. and Mrs. Edward Claus and children, Pete (The Swede) and his brother Axel, Milt Bennet, Olive Dillingham and her daughter Gail, Lillian and Phillip Gregg and their children, Phyllis Jean, Ruth Ellen, and Ronald Stewart.

"The basement of the Medical Center served as the meeting place for Sunday services until over-crowding forced the congregation to seek another place. Through the generosity and understanding of Father Louis Wren, the use of the Holy Cross Parish Hall was offered for the summer months.

"During this time it was indeed evident that the church was growing and was now firmly established, making it feasible to consider a building of their own. After much shopping around, the present building was purchased from the estate of the Amos family on April 4, 1966 and registered as the Beaver Island Christian Church on April 29, 1966. Through volunteer help and donations, a steeple and bell were added, giving the building a new distinction."[41]

Meanwhile, the Holy Cross Catholic Church was completing the job of moving into town. The first Mass in the reassembled Church was held in February, 1958. The next task was the building of a new rectory behind the Church; this was completed in 1961, under the guidance of Father Louis Wren. Father Wren came to the Island in 1959, following Father Alexis Martini, a temporary replacement for Father Herp. Finally, a fund raising drive for a new convent was started in 1966 and the building was completed in 1969. After much work, prayer and financial backing, the task of moving the Holy Cross Parish one and a half miles was finished.

Father Louis Wren served the Parish for eleven years and his departure was a sad event, as he was well-loved by the parishioners. His replacement, Father Herbert Graf served Beaver Island for six years. He was an energetic man who landscaped the church lawns, and had the church insulated, rewired, and reroofed. In 1976, he was replaced by Father Alvin Yard, the present Pastor.

Memories of those who went before us still remain; a white cross at the site of Bishop Baraga's first church; names left by the Mormons — Font Lake, King's Highway, Lake Geneserath, St. James; the Island churches and their furnishings; the ruins of the House of David colony on High Island; and the people of Beaver Island, whose Faith in God was passed down to them by those who went before us — those whose Faith and lives wrote this history.

"God, my God . . . you have made the past, and what is happening now, and what will follow. What is, what will be, you have planned; what has been, you designed." (Jdt. 9:5)

REFERENCES

1. *Michigan Pioneer Society Historical Collections,* II, p. 45

2. "A Short History of the Beaver Islands," *Michigan History,* Annual Meeting 1902, XXXII, p. 178-79

3. Letter of Reverend Frederic Baraga to the Leopoldine Foundation, Vienna, July 1, 1832; copy on file at Holy Cross Rectory, St. James, Michigan

4. Foerster, Fr. Pascal & Soland, Fr. Ewald, "History of Holy Cross Church, Beaver Island, Manitou County (in 1878)," on file at Holy Cross Rectory, St. James, Michigan

5. Cronyn, Margaret & Kenny, John, *The Saga of Beaver Island,* (Braun & Brumfield, 1958) p. 25

6. *Michigan History,* XXII, p. 297

7. *Holy Cross Parish – Beaver Island, 1860-1960*

8. Fitzpatrick, Doyle, *The King Strang Story,* (National Heritage, 1970) p. 51

9. *Ibid.,* p. 70

10. "A Michigan Monarchy," furnished by Charles J. Strang to the *New York Times,* Sept. 3, 1882, *Michigan Pioneer and Historical Collections,* XVIII, p. 631

11. *Ibid.,* p. 632-33

12. "The Murder of King Strang," furnished by O. Poppleton to the *Detroit Free Press,* June 30, 1889, *Michigan Pioneer and Historical Collections,* XVIII, p. 626-27

13. Cronyn & Kenny, *loc. cit.,* p. 61-62

14. Kraus, Henry, "Patchynog's Ailing Island," *Michigan History,* XXXIX, p. 405

15. Malloy, Loretta, Paper written about St. Ignatius, on file at Holy Cross Rectory, St. James, Michigan (1943)

16. *Holy Cross Parish – Beaver Island, 1860-1960*

17. Foerster & Soland, *loc. cit.*

18. Copy of contract between Bishop Baraga and Alexander Guilbeault, July 23, 1860, Burton Historical Collection, Detroit Library; reprinted in the *Detroit Free Press*

19. Kraus, *Michigan History, loc. cit.,* p. 408

20. Foerster & Soland, *loc. cit.*

21. *Holy Cross Parish – Beaver Island, 1860-1960*

22. Foerster & Soland, *loc. cit.*

23. Paper on file at Holy Cross Rectory, St. James, Michigan

24. Kraus, *loc. cit.,* p. 407

25. *Holy Cross Parish – Beaver Island, 1860-1960*

26. Malloy, *loc. cit.*

27. "Dormer Records," Beaver Island Museum

28. Habig, Marion, O.F.M., *Heralds of the King* (Franciscan Press, 1958), p. 556-8

29. *Holy Cross Parish – Beaver Island, 1860-1960*

30. Habig, *loc. cit.*

31. *Holy Cross Parish – Beaver Island, 1860-1960*

32. *South Bend Tribune,* August 3, 1969

33. Davis, Marion Morse, "A Romantic Chain of Islands," *Michigan History,* II, p. 378

34. Malloy, Lawrence, Paper about Garden Island Church, on file at Holy Cross Rectory, St. James, Michigan

35. Cronyn & Kenny, *loc. cit.*, p. 126

36. *Holy Cross Parish – Beaver Island, 1860-1960*

37. *Ibid.*

38. Kraus, *loc. cit.*, p. 408

39. *Holy Cross Parish – Beaver Island, 1860-1906*

40. Vischer, Vivian, "St. James Episcopal Mission," paper submitted for this history

41. Gregg, Phillip, "History of Beaver Island Christian Church," paper submitted for this history

Churches on High Island (top) and Garden Island (bottom).

Left to right: the ice house, the store house, and Sister Dear's house.

The "Rising Sun" at the dock on High Island.

THE HOUSE OF DAVID COMMUNITY ON HIGH ISLAND

by Grant Hart and Phil Gregg

In 1903 a religious sect was started in Benton Harbor, Michigan by Benjamin Purnell, a colorful character with a flaming red beard — "redder than Strang's" — and shoulder length hair. He claimed to be the "Seventh Messenger" referred to in the Book Of Revelations in the Bible.

"King Ben" — a name given him by the papers — was born in the hills of Kentucky in 1861 and had little formal schooling. He became an astute and fiery orator, and with his wife Mary travelled as an itinerant preacher before becoming attached to a House of Israel group in Detroit. When that group's leader, Michael "Prince Mike" Mills, had difficulty with the police over a morals charge, the Purnells made their way through Ohio and Indiana to Benton Harbor.

The sect he founded there believed that the world would soon be shattered by a cataclysm from which only they would emerge, with their bodies intact, to live a thousand years of heavenly life on earth. In the meantime they lived in a communal colony and owned a large and elaborate amusement park.

King Ben soon led his group to prosperity. They worked hard in all the trades. They turned over their income to their king and in return were provided food, clothing, and shelter on a modest level. Some of those joining had considerable means, and cult holdings at one time reportedly included oil wells in Texas. Their investments were made with care and varied from small businesses to Benton Harbor's trolley car system.

In 1912 they purchased land on High Island from the government and established a colony there for lumbering and farming. Most of the wood and produce was sent to Benton Harbor in their own schooner, the "Rising Sun." When that ship was wrecked in 1918, the "Rosabelle" replaced it. Cabins constructed from High Island lumber can still be seen at the House of David in Benton Harbor.

The High Island Community was a thriving colony, with a large dock, a saw mill, a portable shingle mill, corn cribs, an ice house, a laundry, chicken coops, underground storage cellars, a blacksmith shop, a bakery, and many houses. Their fertile gardens yielded bumper crops of cabbages, potatoes, beans, corn, and other vegetables. The Israelites were friendly and hard working, and their relationship with both the Indians on High Island and the Irish on Beaver was good.

This was an old blacksmith house before the Israelites arrived; they converted it into men's dormitories. At the right is an old lumberman's cabin, also built before the House of David arrived.

The main building. A dining room and kitchen were on the ground floor; upstairs there were four bedrooms plus a common room in the center.

Strictly vegetarians, they didn't believe in killing, and even though they lived in the midst of one of the best fishing grounds in the country they refrained from entering that industry. They ate in one large dining hall and the food was prepared by their own specialists. All had their own job to do, and for the most part they did them well. Anything they earned beyond their immediate needs went directly to the mother church in Benton Harbor.

The children of the colony attended the Public School on High Island. One of the teachers, Della Wyland, was an Israelite. Another, Madelyn Kisabe (or Kishego) was an Indian. Among the Beaver Islanders who taught there were Lucille Gillespie, Catherine Floyd, Jane McDonough, and Mabel Connaghan.

King Ben rarely visited his island domain, but instead placed one of his Church Elders in charge, two of whom were William Wright, an architect from Australia, and George Baushke. Mr. Baushke is fondly remembered on Beaver Island. He supervised the frequent trips the Israelites made to Beaver in the boat they'd engineered and designed, the "High Island," to sell their produce. Though considered temperate, he wasn't above bending an elbow with the Irish fishermen in the local bistro, in whose cash registers much of the profits from their vegetables frequently ended up.

Oftentimes in the summer many of the Israelites would take a Sunday outing to Garden Island. They enjoyed playing baseball in the large field along the eastern shore of Indian Harbor. Rich home-made ice cream, cakes, and a wide variety of other food made the picnic complete. At the end of the day they would board their boat and amid laughter and singing head back for High Island.

The large eight-sided building that was first presumed to be King Ben's castle was used to house the young Israelite girls until they were of an age to be married — which was an unfulfilled institution for the House of David. In the meantime the girls' activity was restricted to the working day; at night guards were posted outside the "House of Virgins."

Logs ready for shipping.

In the winter the industrial activity on High Island was confined to cutting timber. When the ice thickened transportation between Beaver and High was easy for horse drawn cutters and ice sailing boats. On one occasion Beaver Islanders paying a visit by auto were flabbergasted to discover, when it came time to leave, that the ice had gone from between the islands. They watched helplessly as their car drifted away on a cake of ice.

Marion Morse Davis, in his article, "A Romantic Chain of Islands," quotes Arthur Stace, who visited High Island in 1926: "The little school on High Island is attended by Indian

and House of David children. A white church steeple guides the way to the settlement. This church is a Catholic Indian Mission attended by a Franciscan missionary from Harbor Springs. A dock juts out into the lake. Behind it is a sawmill. A large rambling boarding house has several dozen one-story cabins scattered behind it, mostly frame structures with shingle sides. All the men, from patriarch to boy in his teens, wear the typical House of David cap, baggy because it must contain a lifetime growth of hair. One of their tenets is to go unshaved and unshorn. Within the little cabin yards, each closed in with a high fence of poles, are splashes of summer beauty — blooming flowers. The timber has been fairly well cleared away from the island so the sawmill is not running regularly; the main occupation now is farming. Their belief is that they are a chosen people who will live when others have perished."

On Beaver Island it was said that because of this belief a death was taken as a sign the person's faith had lapsed; the person was posthumously excommunicated and his body was taken to another island for burial.

In October of 1921 the "Rosabelle" left High Island with a load of lumber and 600 bushels of potatoes, but she never reached her destination. The hull was discovered a week later by a Grand Trunk Ferry, floating bottom up with her sails and rigging dragging along. No survivors were found.

The "Rosabelle"

On December 16, 1927 Benjamin Purnell disillusioned his flock by dying. An order that the sect be disbanded was dropped and Judge Dewhirst, an Israelite official from Benton Harbor, took over the leadership of the colony. A fight ensued for control of the multi-million dollar kingdom between the judge and Mrs. Purnell, "Queen Mary." In 1930 a settlement was reached in court and Queen Mary left the House of David with about 200 supporters.

The judge gained control of the High Island venture, yet the colony was anti-Dewhirst and pro-Mary, causing the end of their community. The judge sent William Wright to the island to supervise the evacuation and to take inventory of his assets.

Some enterprising Beaver Island traders foresaw a golden opportunity. The Israelites had raised some beautiful livestock, milk cows and prime breeding stock. The Beaver Islanders had a number of decrepit animals that were nearly ready for the fox farm. While Billy Wright waited for winter to end to ship his cattle off the island the Beaver traders brought their feeble stock across the ice and did some secretive swapping. When the ice broke Billy arrived. The right number of cattle were there, but they hardly seemed worth shipping down state. After stroking his beard he finally decided it had been a very rough winter.

The High Island property was sold in the 1950's to Warren Townsend, who wanted to raise cattle there, but the House of David still exists in Benton Harbor. There are only a few elderly members left, since celibacy was one of their rules. The once popular amusement park stands largely abandoned. A visit with the group's present leader, Tom Dewhirst, and his wife Ola Mae (Boone), gave a new perspective to the High Island community. Ola Mae lived there as a girl, and her husband visited the island with his father. They shared their memories and photographs of the island. Mr. Dewhirst recalled that at one time there were over a hundred of their members there, along with a comparable number of Indians who were divided into twelve large families. He remembered that the mill had burned out twice. His wife reminisced about her childhood on the island: her Indian friends and schoolmates — Chief Kenwabikise and "Little Joe"; the Beaver Island residents who came over to put out the Abbe home fire; a cousin who ran away to the mainland; and her name carved on a tree on the highest hill on the island.

That's all that remains of this utopian experiment . . . memories.

U.S. Mail. Bass I.

BEAVER TALES:
THE MAIL CONTRACT

by Phil Gregg

(These 'Beaver Tales' have been appearing with increasing frequency since 1966 in the "Beaver Beacon," Beaver Island's monthly paper.)

BRINGING THE MAIL OVER THE ICE

When winter's frigid breath coats the waters surrounding the islands with a layer of ice, tales of yesteryear and the uses that were made of this phenomenon make today's routine seem pretty plush.

Most notable of the ice travellers are the men who made up the list of Beaver Island's mail carriers, in the days when these men were the winter link with the mainland. Adventure was merely a way of life, and to these men a trip from Beaver Island to Cross Village and back was all in a day's work. For safety's sake this job was most always handled by a team of two men with a horse and a sleigh. The era of the ice carriers began in the late 1800's and lasted through 1930.

The names of Paul LaBlance, Frank and Carl Left, Willie (Brutz) Boyle, Joe Alphonse, Harlem Gallagher, George Williams, Jack Anthony, Raymond McDonald, Charlie Gallagher, Willie Gibson, Dan and Frank Cornstalk, Joe Floyd, Joe (M'Fro) Sendenberg, Charlie Martin, and others who filled in for short periods, make up the list. Around each man many stories hover of harrowing experiences and bitter battles with the winter elements of the Northern Lake.

Winters of the past seemed more severe and the ice more treacherous. Caution was the byword in crossing to the mainland. Between islands the currents always keep the ice thin in places, and visibility was frequently poor, so problems came up which made these trips into perilous adventures.

Almost unbelievable is the tale of the endurance of Dan Cornstalk, a High Island Indian. On several occasions he picked up the mail at Beaver Island and headed out over the ice on foot. Though the sack wasn't heavy, since parcel post hadn't been invented yet, his destination wasn't Cross Village but Mackinaw City. There's a mountain of difference between walking on a sidewalk in the middle of May and walking across the icy wastes of Lake Michigan in mid-winter. Pressure ridges of ice must either be climbed over or walked around. Solid ice isn't always in a straight line from point to point. Drifting snow, crusty or soft, makes the miles extremely long. Dan made the trip to Mackinaw City alright, and the day was but half over so he picked up the return mail and headed out again. Before dark he made it back, not just to St. James but to his home on High Island. For a man in good shape, a hike only to High Island and back would be considered a mighty good day's outing.

Average mail trips to Cross Village with horses and sleigh took the better part of a day, with a stop at Hog Island to feed and rest the horse. Many trips were uneventful and pleasant, with many a good time in Cross Village before returning the next day. There were occasions when the trip back was left entirely up to the horse, since the carriers had met so many friends the night before they merely covered themselves up in the sleigh and went to sleep, oftentimes not waking until the horse stopped at the Post Office in St. James.

Ice conditions can change in a very short period of time, due to wind and temperature, making some trips that start out in fine shape end up in near catastrophe. On a return trip Carl Left and Joe Floyd were making good time with the wind on their backs when suddenly their sleigh lurched up under them, forcing them to jump clear. A pressure ridge had caught the sleigh, mail and all, in a cataclysm of crushing ice. The men watched in awe as the rapidly growing ridge ground the sleigh to splinters, and the mail along with it. The following day they returned to look for the mail. But the ridge of ice, which had towered over thirty feet, had disappeared: no ridge, no ice, and alas! No mail bag.

Frank Nackerman vividly recalls his first trip in 1928, with George Williams and Jim Washagesic. With the temperature hanging at a brisk sixteen below, Frank lost contact with his toes, and his nose and cheeks felt frostbitten too. The ice between Hog Island Reef and Hog Island forced Frank and Jim to walk ahead of the sled to sound for solid ice, while George led the horse. George lost his footing on a piece of heaved ice and fell hard, cracking the back of his head. He got right up, so Frank and Jim failed to realize he'd knocked himself nearly senseless. Finding a safe route around some bad ice, they motioned George to follow.

Not comprehending their motion, George drove the horse onto the weak ice and went through. Frank and Jim were horrified when they looked back and saw George waist deep in water still trying to drive the horse and sleigh. It took some doing, but soon all were on solid ice again. George's clothes quickly froze into a suit as binding as cast iron. When they finally reached home they began the long and painful process of thawing out. George had frozen his toes so badly, though, that they had to be amputated, and the trip ultimately caused his death. Regardless of the dangers and discomforts, these men liked the work and met the challenge with a grim pride as they trod a most unusual mail route.

(March, 1966)

BEAVER ISLAND'S FERRIES

Ferry service to Beaver Island has had a long and adventurous past. Operating a scheduled service over such a wide body of water, offering the passengers comfort, safety, and speed, and making a profit has been a challenge to skippers and boat owners alike. To island residents the ferry has been the most important link to the mainland and the object of special fondness. The advent of the airplane reduced dependence on the boats, but the Island community couldn't survive without them.

Each vessel commissioned to the Beaver Island run has left its legend. The first was Cundy Gallagher, who had the original mail contract and sailed regularly between Beaver and Cross Village. The "White Dan" Green sailed a different vessel on the route, handling the job until 1890 when the first steam vessel was employed, the "Nellie," skippered by Captain Peter Campbell. He replaced the "Nellie" with the "Erie L. Hackley" and operated this ship until he purchased the "Beaver" in 1902. The "Beaver" caught fire and burned at the dock in Charlevoix, and a dilemma developed until another vessel could be found.

The Roe Brothers, Jim and Charlie, of Harbor Springs, patched the break when they picked up the mail contract and operated their "City of Boyne." After a few years they replaced her with the "Columbia." It was obvious that small boat owners weren't vying for the run. No vessel had yet been designed expressly for it. Profits weren't great. And several of St. James' fishing tugs frequently carried freight and passengers across, so there was no monopoly.

(February, 1975)

Thanks to some letters and old newspaper articles we've uncovered some additional facts about the ferry run.

ISLAND FERRIES, CONT.

John A. Gallagher wrote us about "Big Neil" Gallagher's boat, the "Joe." "Big Neil," who had a store on the island, a fish business, and a tug named "Gertie," used the "Joe" on the run in 1900. Built in 1889, she had an 80 × 16 keel beam, a gross tonnage of 56, and a net of 52. Skippered first by Captain Leonard and then by James Gallagher, she travelled a route that included points in Little Traverse Bay and Cross Village; in those days the few roads in existence were rutty strips of mud or choking dust, so travelling by boat was more common than it is now.

After two seasons the "Joe" was replaced by the "Beaver." When the "Beaver" burned the "City of Boyne" was given the contract.

The "City of Boyne" was a fairly good sized vessel, measuring 87.8 feet overall with a beam of 19.7 feet and a draw of 7.8 feet. Her gross tonnage was 121, her net 82. She carried a crew of five. Her steam engine developed 45 horsepower.

Originally named the "May Blossom," she was built of wood in Grand Haven in 1883. She carried passengers up and down the west shore of Lake Michigan from Grand Haven until 1900, when she was purchased by Oscar E. Wilbur of Charlevoix. She was used until 1914 on two daily runs from Charlevoix to Boyne City, and became a familiar sight in such ports as Horton Bay, East Jordan, and Petoskey. Sports fans used her to follow their team, and even the sheriff transported his guests to the County Jail from the brawling waterfront dives that were so prevalent 75 years ago.

(March, 1975)

THE "HACKLEY"

The "Erie L. Hackley" was one of the first island ferries, and has a history about which there is much controversy.

She was built in 1882 in Muskegon for Captain Seth Lee's Muskegon Lake Ferry Line. Her dimensions were 79 × 17.4 with a maximum draft of 5.2 feet, a gross tonnage of 90, a net of 75. She served Captain Lee, along with the "Centennial" and the "Mary E. Menton" until he sold out in 1893.

Then Captain Peter D. Campbell acquired her and operated her between Muskegon and Whitehall until 1898. In 1899 Captain Campbell moved to Charlevoix and put his steamer on the run from Charlevoix to the Manitou Islands, according to Mrs. Fred Floyd of Muskegon and Janet Coe Sanborn, Curator of the Cleveland Picture Collection and Editor of "Inland Seas."

Other sources claim she ran from Charlevoix to Beaver Island for almost ten years.

In 1902 Captain Campbell sold her to Benjamin Newhall of Chicago, but remained as Captain. But the next year she was sold again, to the Fish Creek Transportation Company of Menominee, Michigan, for service on Green Bay. During the summer of 1903 she ran to various ports on the Door Peninsula under the command of Captain Joseph Vorous of Fish Creek, who, with Henry Robertory, was the principal owner.

On October 3, 1903, the "Erie L. Hackley" cleared Menominee at 5:05 p.m. and set a course for her first stop at Egg Harbor. Perhaps he figured the wind would die down as the day ended, but Captain Vorous soon found his vessel caught in a real blow. At 5:57 she was in trouble off Green Island. She broached-to several times and finally fell into a trough and was swamped by the huge waves. She was carrying 12 passengers and a crew of seven. In a matter of moments the vessel went down. Survivors took refuge on floating wreckage in the raging and bone-freezing water. One group of six floated for 14 hours on a deck covering, until they were picked up by the Goodrich steamer "Sheboygan." Two others were saved, but 11 lost their lives. Captain Vorous was last seen in the pilot house desperately trying to head the little vessel into the wind. The foundering is annually remembered in Menominee as the city's worst marine disaster.

Some days later the cabin of the lost ship floated ashore near Eagle Bluff. All life belts were found in place, attesting to the suddenness of the ship's sinking. The "Hackley" was gone but not forgotten, as the depths of Lake Michigan hold her remains.

(May, 1973)

THE "BEAVER"

Probably Captain Campbell's feeling that the steamer "Hackley" was too small to battle the wiles of Lake Michigan prompted him to sell her after such a short time, and perhaps her loss the next year indicated he was right.

The Booth Packing Company of Chicago had a boat for sale that seemed to meet his requirements. Built in Grand Haven in 1892, she was given the descriptive name "Oval Agitator." She was 98 feet long and only 19 feet in breadth,

137

and drew 9 feet at 121 gross tons and 80 net; her name was appropriate. Captain Campbell purchased her in 1902 and renamed her the "Beaver." She'd been built as a fishing tug, but after some modifications she began a 12 year run to Beaver Island.

Captain Campbell's son, P. D. Campbell, became her skipper in 1913. In the summer of 1915, while lying at the dock in Charlevoix, a fire broke out on the "Beaver" and before it could be brought under control all her upper works were burned off down to the car deck. Repairing her wasn't feasible so she was shorn of her dignity and towed to St. James, where her bare hull was used as a barge. Her heyday was over and the gaunt remains of the "Beaver" faded into obscurity. In 1930 she was classed as a "tow," owned by Henry Martin of Menominee, Michigan; her home port was Milwaukee.

In 1962 the Flint Treasure Map Enterprises published a map listing a vessel "Beaver" as having sunk in Sturgeon Bay; no date or other information was given.

This sailing scheduled for the "Beaver" was in an ad posted in Charlevoix:

"Steamer Beaver — P.D. Campbell, Master and Owner

Three round trips weekly to Beaver Island leaving Charlevoix in the morning, Monday, Wednesday, and Friday, arriving at the island at 11:00 a.m. Returning, leaving the island at 1:30 p.m. and arriving at Charlevoix 5:30 p.m.

Tuesday, Thursday, and Saturday can be chartered for excursions."

Those who are familiar with Charlevoix may remember the Campbell Boot Shop that operated there until recent years. Owner and operator: P. D. Campbell, ex-skipper of the steamer "Beaver."

(June, 1973)

THE "COLUMBIA"

The mid 1900's saw several vessels of various sizes and descriptions serving the Beaver Island run, some of which had very short terms of service, such as the "Columbia." She looked very much like the ship she replaced, the "City of Boyne." Her skipper was Charlie Roe of Harbor Springs and her engineer Pat O'Donnell of Beaver Island. Her career lasted until early October of 1917.

Being a steamer, it was imperative to keep a fire in her boiler while she laid over during the night at her dock in Charlevoix. Pat O'Donnel pulled out the clinkers, shoveled some coal in, banked the fire, closed the draft, and 'bunked down' for the night. Apparently a red-hot clinker fell into the oil-soaked bilge, for Pat was awakened during the night by the smell of smoke and the crackling of flames, which had engulfed the engine room and cut off his access to the gangway. Not having time to consider modesty, Pat made his way to the foredeck in his longjohns. Unable to swim, his only escape was to slide down the bowline to the dock. As the flames lit up the chilly October sky Pat's dilemma was also illuminated and someone offered him an overcoat to ward off the cold. Bucket brigades and hand pumpers could do nothing to stem the fury of the burning ship. The dock and coal sheds of the Charlevoix Coal and Wood Company also caught fire and burned as the "Columbia's" hissing and steaming hulk slid beneath the water, closing a short chapter in the long list of boats serving Beaver Island.

(April, 1975)

THE "BRUCE"

With the disasterous loss of the "Columbia" in 1917, Captain Harry L. Oldham put the screw steamer "Bruce" on the Beaver Island run. A vessel of several names, she was built in

THE "BRUCE," CONT.

1892 in Chicago. She was 66 gross tons, 44 net, measured 78.7 by 19.9 with a draw of 8.5 feet. She could develop 225 horsepower.

She began her career as the "A. F. Yattah," a passenger carrier from Grand Haven. Then she was renamed the "Ogontz." Captain Oldham brought her to Charlevoix and renamed her again.

From 1917 to 1920 she had a competitor, the "Irene." The "Irene" was smaller, slower, and certainly didn't offer as many creature comforts. She soon faded into the past, leaving the "Bruce" to handle the load through the season of 1920.

Many remember Harry Oldham as the builder of the old Tower Hotel and Casino, overlooking the entrance to the Charlevoix channel.

(May-June, 1975)

THE "SANFORD"

In 1920 Captain Oldham retired his steamer, the "Bruce," from the run. Later in the year Captain James Sanford purchased a 100 foot boat he named after himself and used on the Beaver Island run for the next ten years.

The "James E. Sanford," which was to be the Beaver Island boat for the next ten years, was a coal burning, wood-hulled steamer that offered its passenger comparative comfort on the 5½ hour run.

The time had now come when there was a demand for transporting vehicles, and the "Sanford" was pressed into service to meet that demand. Being open at the stern, she could accommodate one car, providing the lake was cooperative. There were restrictions, however, in that the overhead clearance was limited and large sedans had to be crossed off the list, unless they happened to be the popular touring model with a fabric top that could be taken down. Loading and unloading "horseless carriages" was a real project which always attracted a spectator crowd and made the car owner nervous.

In 1925 Captain Sanford retired from the run and turned command of his vessel over to Captain John Chambers, who commanded her until the end of the 1930 season. A man who kept the pilothouse pretty much to himself, perhaps because of his fondness for limburger cheese, Captain Chambers handled his job with dispatch until 1931, when demand called for a larger and faster vessel for the Beaver Island run.

(November, 1975)

140

THE "OSSIAN BEDELL"

Beaver Island was one of the few places in the nation not to be severely affected by the Great Depression. The islanders, with their own thriving industries of commercial fishing and lumbering, plus their productive farms, found themselves pretty independent. This was an era of keen competition for the Beaver Island run.

When the "Sanford" was retired the gap was quickly filled by Captain James ("Big Neil") Gallagher with the "Ossian Bedell," a bigger and quicker ship. Her main deck was large enough to accommodate several cars at a time, and her large promenade deck gave the passengers plenty of elbow room. Her stay on the run was rather short-lived, however, as Captain Gallagher found a more lucrative run for her in 1933. This was the time of the World's Fair in Chicago. The demand was great for moving multitudes of people from the city of Chicago up the Chicago River to the fair grounds. Captain Gallagher took the "Ossian Bedell" down the lake and established her on the new run, leaving the Beaver Island run up for grabs again.

After years of smaller, slower boats, it was a real treat for the islanders to have a vessel like the "Bedell." It inspired one of the largest Homecoming celebrations ever in 1932. It was indeed disappointing when she left, for the islanders had had a taste of comparative luxury they would never have again.

THE "OSSIAN BEDELL," CONT.

To fill the gap temporarily, the little Coaster Steamer Gus Mielke had built, the "Rambler," was pressed into service. It was at this time in 1933 that the ill-fated "Marold II" came on the scene to write the most dramatic chapter in the long line of boats to Beaver Island.

(February, 1976)

THE "MAROLD II"

When the "Ossian Bedell" left the area Captain Sanford brought back the "James E. Sanford" to fill the void until June, 1933, when the "Marold II" arrived under the command of Captain Ludlow L. Hill, with his son Leon as the engineer. She was to write the most harrowing chapter among all the stories that accompanied the many boats to Beaver Island.

She began her career as the ultimate in private yachts. She was designed by Cox and Stevens, Naval Architects, and built by John H. Dialogue and Sons of Camden, New Jersey for Commodore Alexander Winton, who manufactured the Winton Auto. She was launched in 1911 and christened the "LaBelle." Perhaps the most beautiful and luxurious steel yacht of her day, her statistics would still establish her among the very upper crust of Great Lakes Yachts.

She measured 145 feet long with a graceful clipper bow and a fantail stern. She was narrow of beam (19.1 feet), but carried several tons of lead ballast plus three six cylinder 175 horsepower gasoline engines. She was stabilized like a sailing vessel but had the power to move at an impressive 13 miles per hour. Apparently, though, this didn't impress the Commodore, for he soon replaced the engines with a pair of twelve cylinder V-type Winton diesels of his own design. This increased the speed to 15 miles per hour. Satisfied, the Commodore made her the flagship of the Inter Lake, and she cruised the Great Lakes from Lake Erie to Lake Superior.

In 1919, the by then well known "LaBelle" was sold to Harold Wills, a manufacturer of the Wills-St. Clair automobile of Marysville, Michigan, along the St. Clair River. Mr. Wills combined his and his wife's name, Mary and Harold, and renamed her the "Marold II" — it was his second yacht.

Just two years later she was destroyed by fire and settled to the bottom at her dock in Marysville, ending the luxurious life of the graceful yacht but not ending her career.

(March, 1976)

The gutted but sturdy steel hull of the "Marold II" was raised from the St. Clair River bottom in 1925, four years after she'd sunk, to begin a much different role.

The Hill-White Transit Company purchased the vessel and proceeded to convert her into a combination passenger and freight ship. Her once gleaming white hull was painted black, one of her diesel engines was removed, and the other placed amidships, making her a single screw ship. The spare engine was used for parts. Eventually all the spare parts were used, and she was repowered once again, this time with a Fairbanks-Morse diesel.

In the course of conversion the frills that identified her as a yacht were removed. The bow-sprit was taken off and her main deck was enclosed to accommodate freight and passengers. A pilot house was put on her superstructure and an exhaust stack, hardly befitting a yacht, was placed aft of the pilot house. Additional passenger accommodations were added to the top of the superstructure. She operated on Lake Michigan in a number of trades under first the Hill-White Transit Company and then, when the Hill family bought the White interests, the Hill Transit Company.

In September, 1935 the "Marold II" was purchased by the newly formed Beaver Island Transit Company, and it looked like at last Beaver Islanders were going to have a boat of their own for the Charlevoix run. The company consisted of Everett Cole, Hugh Roland, Bruce McDonough, Lloyd McDonough, and Nels J. LaFreniere. Captain Hill and his son were retained to sail on her until the end of 1936.

(April, 1976)

The fate of the "Marold II" became entwined with the fate of another ship, the "J. Oswald Boyd," a small U.S. tanker which transported petroleum products first on the Atlantic and then on the Great Lakes.

The "Boyd" was built of steel in Scotland in 1913, and measured 244 feet long, 44 feet in beam, and 20 feet in depth, with a gross tonnage of 1806 and a net of 1213. In 1923 she was bought by a New York company, and in 1929 she changed hands again.

On November 11, 1936 she encountered a terrific gale in northern Lake Michigan and was driven aground on Simmon's Reef, stern first — a total loss. Her cargo was a load of gasoline.

When the "Boyd" was abandoned, many people from Beaver Island and Charlevoix started removing the gasoline and selling for as little as pennies a gallon. It was stored in every kind of container: drums, cans, bottles, even open pails.

At first the "Rambler," a 65 foot long package freighter, was employed to transport the gasoline off the "Boyd," starting almost the day she was stranded. After she'd completed 24 trips it was decided to use the "Marold II," which had a capacity of 17,000 gallons as opposed to 9,000 for the "Rambler."

On January 1, 1937 the "Marold II" cleared Charlevoix and headed for Simmon's Reef for another load of the gasoline that was there just for the taking. It was a beautiful sunny day, so perfect that Ludlow Hill and his son Leon, Everett Cole and his brother Raymond, and Bruce McDonough decided to forsake the holiday.

The "Marold II" usually left Simmon's Reef around 5:00 p.m. for her return voyage to Charlevoix. At 5:30 p.m. the lookout at the Charlevoix Coast Guard Station sighted a column of black smoke in the direction of the reef. A lifeboat was dispatched, and arrived at the reef, 40 miles away, about 11:30 that night. Other boats were already there from closer ports. It was dark, but a raging fire lit up the sky. The rescuers were helpless. When morning finally came the degree of the tragedy was revealed.

The explosion had peeled off the main deck and superstructure, the pilot house, and the stack of the "Marold II," hurling the debris onto the deck of the "Boyd" in one tangled mass. Soon the remains of the "Marold II" slipped

144

below the water, her hull having been blown apart. There were no survivors.

When the metal cooled on the "Boyd" a boarding party found the bodies of the Cole brothers, identified through their dental work. The following spring Captain Hill's body washed up on the southwest end of Beaver Island, and his son's body was found near Fairport, near the entrance to Green Bay on Lake Michigan's western shore. Bruce McDonough's body was never found. The insurance on the vessel had expired at midnight, December 31, 1936, and had not been renewed.

For years after she sank her clipper bow could be seen in about 22 feet of water. The hulk of the "Boyd" was pulled off the reef and scrapped when the war raised the demand for steel.

(Not yet published.)

In 1937 Leroy Allers began operating the "Mary Margaret" and Jack Lyons the "North Shore." It took some time for the Beaver Island Transit Company to recover from the loss of the "Marold II," but when they did they took over the "North Shore." In 1955 the Beaver Island Boat Co. bought the "Emerald Isle," and in 1962 they added this fine ship, the "Beaver Islander."

One of the many Beaver Island Bands
(left to right: Cap Ludwig, Gus Mielke, Herman Allers, George Burrell, Cap Allers, Erwin Belfy, Charles Allers, Neddeau, and George Herrick).

IT HAPPENED ON BEAVER ISLAND

by Margaret Hanley

(These are excerpts from Mrs. Hanley's newspaper column, "It happened on Beaver Island," which appeared in the Charlevoix Courier between 1947 and 1951.)

It is always a thrill to see the first boat to the Island after three months without navigation. Saturday, when the White Swan brought over the mail, only a few drove down to the south end of the Island where the Swan lay anchored a mile off shore. Unable to get into the harbor, the Swan had to be unloaded at the head. Only mail was brought and one passenger, John Ricksgers.

(April 13, 1949)

The White Swan came to the Island loaded with coal, Carl Kuebler is the sales manager and distributor of the "black gold."

(August 27, 1947)

Along with the wind storm which swept the Island the first of the week came the long-looked for snow. Freezing temperature made the snow ideal for fun on the Hall Hill — the one spot in the village where children can enjoy a long, speedy sled ride. Many of us who live in the vicinity of McDonough's store, where the gusts of wind hit the hardest, felt as though the roofs over our heads would lift off at any moment.

Art Miller's boat, anchored about 100 feet off shore with a washtub of hardened cement, was blown to shore — anchor and all. Several trees around the town were blown down but no serious damage was done.

(December 15, 1948)

Monday afternoon, the city of St. James became as boisterous as I have ever seen it. All housewives ran to the windows, children ran to see and the town was in an uproar. The reason: Coming into the harbor was the John Roen IV, giving salute after salute. Each time, either the Rambler of Gus's mill would respond and the tootin' rootin' surely gave us something to talk about. John Roen, owner of the boat, was at one time a partner of Gus Mielke.

(August 6, 1947)

147

Three sets of teeth have been lost during the past year. Two sets in the Jordan River and one in the big lake. Any day now, the fishermen expect to lift a big lake trout sporting a glistening set of ivories.

(April 7, 1948)

The harbor is completely free of ice and once more we can feast our eyes upon the blue waters of Lake Michigan. Recipe for de-icing a harbor: one trip by the C. G. tender, Mesquite, one generous cutting by Pickett boat, twelve hours of strong Nor-west wind. Recipe tried and tested on Beaver Island. The results are perfect.

(April 14, 1948)

A mixture of snow and rain and freezing temperature iced up the island Sunday. Slippery roads kept a lot of folks in, but the young set were at Bundy's Hill having the time of their lives coasting. Strong winds have taken away the majority of the ice in the harbor and again navigation is open to the dock.

(Jan. 19, 1950)

The Ice Breaker "Woodbine" came to the island on Thursday to cut a path to the dock. The Woodbine made its appearance about 7 a.m. and it was close to noon before it reached the dock. The ice in the harbor looked to be 20 inches thick. Thursday night, nature's icebreaker, the sou'west wind, blew up enough strength to break up and take away the ice outside the harbor and to take away a number of fish shanties too. A few shanties remain inside the harbor. The team of Belfy and Boyles have their bright aluminum shanty in a good spot. One day last week, their nets yielded 190 pounds of perch.

(Undated)

* * *

The fish shanties are on the increase but the perch are not. Fishermen shift their shanties around on the harbor trying to find a likely spot but as yet the catch has been skimpy.

(Feb. 25, 1948)

Fishermen's nets have been heavy with perch caught in the harbor. As much as 600 pounds of perch have been brought in with one lift.

(October 5, 1949)

Shanty Town boomed to seventeen shacks this past week. The weight of the village on ice is on the southwest side of the harbor at the present. Fishermen who are lowering nets beneath the ice are making good hauls. Alec Cornstalk, and his son Franklin, brought in 120 pounds Saturday afternoon.

(February 21, 1950)

This is no fish story — Culls strung out 30 miles of nets the first week out on the lake.

(April 21, 1948)

Stanley Floyd and Dick Martin, aboard the "Helen R.," braved the rough seas Thursday to lift nets. With the job of lifting over, engine trouble developed in the vicinity of Garden Island. To add fuel to the fire, they lost their anchor. Without food, the men were out in the lake with the nets lowered to anchor the boat until Saturday. The Coast Guard went to their rescue. It was an all day job for the picket boat to pull the "Helen R." into St. James harbor due to the roughness of the lake.

(November 10, 1948)

The tug "LaBlanc," which went aground on Beaver Island on December 5, was pulled off the shore near Beaver Head lighthouse at 11:15 Thursday morning by the Coast Guard cutter "Woodbine." Working at a distance of 3500 ft. from shore, the "Woodbine" under the command of Lt. V. H. Leech, Jr. ran an eight inch line to the tug which had been taken high and dry on the beach by the heavy sea and high wind. After more than three hours of maneuvering and with the aid of the Beaver Island lifeboat crew, the tug was hauled off the beach and the reefs.

(December 15, 1948)

Cull's fish tug, "LaFond" went out of the harbor early Monday morning with eight boxes of nets to set. During the week fourteen of the boxes — equalling five miles of nets were lifted. The yield again showed the effects of the destructive eel. Listed were 125 lbs. of fish and only one trout in the whole bunch.

(April 27, 1950)

* * *

Gus Mielke started the mill running Tuesday morning. First off came the hemlock boards that will be used to repair LaFreniere's dock.

(April 21, 1948)

149

Mielke's mill was forced to close down Monday due to the breakdown of a vital part of the engine. The engine, made 85 years ago, was formerly owned by the Israelites and was bought twenty years ago by Mielke and moved to the Island from High Island. It is hoped the part can be repaired and work at the mill will be resumed in a short time.

(Oct. 5, 1949)

The North Shore was loaded twice this week with Beaver Island hardwood, the beginning of moving around 300,000 feet of lumber for delivery on the mainland.

(Sept. 21, 1949)

The White Swan, which was loaded Wednesday and Thursday with 75,000 ft. of Wojan's hardwood, made two attempts to cross the lake, but Lake Michigan has been on a rampage, and the Swan came back to roost at LaFreniere's dock. Sunday morning the Swan was gone.

(Undated-1947)

* * *

Mettie Perron, a Frenchman who has lived on the island for many years is a champion berry picker. During strawberry season he picked 125 quarts, and it takes a lot of picking to make one quart of the wild berries. Mettie is busy picking raspberries now.

(July 2, 1947)

Beaver Island's gardens are looking good. With very few scorching days and plenty of rainfall the harvest is gratifying. Farmers are bringing beans, beets, cabbage, peas, and cucumbers to market. There seems to be no end to the amount of raspberries and huckleberries one can pick this summer.

(August 25, 1948)

It's cranberry time on the Island. Groups going out to the marshes are getting a nice supply for the winter. Blackberries are very scarce. The long dry spell caused the berries to drop off the bushes before ripening.

(September 7, 1949)

Mr. and Mrs. David Wilson invested in an electric-powered cider press to run through the twenty-two gunny sacks of apples gathered in their orchard. Cider making started in earnest Tuesday, and continued on through Wednesday; the results, a barrel of cider for the Wilsons to take home and thirty-eight gallons donated for the Hallowe'en party to be given at the Parish Hall on Hallowe'en night. Thirteen additional sacks of apples were gathered from Charlie "Harlem" Gallagher's orchard to complete the job for the party.

(November 2, 1950)

Earle Boyle's canning spree has run him out of containers. Apple sauce to Earle is like spinach to Pop-Eye, so with the bushels of apples yet to be canned, don't throw away that empty jar — save it for Earle. By the way, that bit of notoriety Earle rated in the Courier has produced a bit of enthusiasm among the fairer sex. One anxious lady made a special trip from Grand Rapids to meet the Island's Canning King. There's no telling where this apple sauce is going to lead — maybe to the altar, huh!

(October 13, 1949)

Beaver Island maple syrup is so delicious! Two ambitious teenagers Mary and and Robert Bonner, tapped around 100 trees and collected enough sap to yield 14 gallons of maple syrup.

(April 20, 1949)

Beaver Island farmers have formed a cooperative under the leadership of George Ricksgers. The farmers' co-op have purchased a threshing machine and have already put it to good use. The farmers hope to make great strides working and pulling together.

(September 21, 1950)

*   *   *

The largest single haul of Beaver beef this season went across Monday on the "North Shore." Thirty head of cattle belonging to James W. Gallagher went aboard. The boat, with its cargo of livestock, had a rough crossing. Strong wind out of the sou'west gave Captain Dick Lyons a battle, but after five hours of struggle, the "North Shore" safely tied up in Charlevoix. From reports, not a single animal was injured.

(October 20, 1949)

It's round up time on the Island and Beaver beef is going to market. Last Monday, 35 two-year old calves belonging to James W. Gallagher were put aboard the "North Shore" for

151

shipment. Trucking cattle to the dock from the country and getting them aboard the boat is no small job. The majority of the bovine beauties resent being transferred from green pastures to the truck and to the boat and it takes experienced Beaver Island cowboys to handle the situation.

(October 19, 1950)

Sunday, seven cattle were discovered dead, one in Egg Lake and six in Green Lake. The cattle in an effort to reach water had sunk in the deep mire surrounding the lakes. Unable to free themselves, all perished.

(Sept. 22, 1948)

* * *

Small game hunting has excellent possibilities on Beaver Island this season. The woods are full of rabbits and squirrels. Ducks and pheasants are plentiful too. October 15, the opening day for small game, should be an exciting day on the Island.

(September 21, 1949)

Dick LaFreniere does his good turn daily. He keeps a sack of feed in the back of his truck and every day takes a run out to the field opposite the dairy to throw food to the pheasants.

(Feb. 21, 1950)

Roland McCann and Joe Gallagher have taken fishermen, who have come from all parts of the Mid-West, to Garden Island since bass season opened. Bass fishing is excellent in the Garden Island area and all fishermen report catching their limit.

(July 13, 1949)

Beaver Island — First sharp-tail grouse hunting ever allowed by the conservation department here attracted 16 hunters who shot 40 birds. Season closed November 5. This was an average of two and one-half birds each. Species was introduced on this Charlevoix County island ten years ago and the original colony of 29 birds is estimated now to have increased to over 200.

(November 16, 1950)

Beaver Islanders are preparing for the invasion of a number of deer hunters from the mainland. Due to getting cars across, many hunters come to the island two or three days in advance. This gives them time for a lot of visiting and "shop talk" about who's going to get the biggest buck. Of course not all the hunters get a buck and that explains the row of one dollar bills tacked on the wall by the side of the Shamrock Bar. Many hunters who do not get a buck to take home leave a "buck."

(November 9, 1950)

Island trappers spent a busy week luring beavers into their traps. The season did not look too promising due to the frozen streams, but thawing weather towards the end of the week brought profitable results. Chuck Kleinheintz has the high score so far. He brought home four beavers Friday night. Buddy and Bruce McDonough have two to their credit. The season has another week to run. Beside the bounty collected for the pelts, the successful trappers will have a fine feast of roast beaver. They say the meat is delicious.

Bub Burke dropped a coyote Sunday afternoon in the woods back of Carl Keubler's home. "Well, I never had such a shaking up since I broke in a wild horse to saddle," said Old Timer after a ride over the foot deep muddy ruts on King's Highway Sunday. More than one car had to be pulled out of the mud going back and forth to church Sunday. Oh well, it's spring, so what's a little mud!

(March 30, 1949)

The grand total after Beaver season closed was 16 beavers trapped. Mr. and Mrs. Chuck Kleinheintz trapped five each and the McDonough boys three each. The beavers range in size from 19 to 43 pounds.

(April 13, 1949)

The annual hunting expedition to Sand Bay was a memorable one this season; the fellows actually brought back something to show for their efforts. It's a brush wolf measuring five feet four inches long and the critter is on display in the Shamrock Tavern. The story goes that Archie LaFreniere was planted in a clump of bushes while Bruce McDonough and Donald Cole on horse back circled the area for tracks. In the wide open field, Archie froze his fingers to the trigger until the wolf was about 10 feet away, then he let 'er go and the wolf was dropped with one shot.

(March 1, 1951)

153

Saturday afternoon, Charles Kleinheintz, and Bob McCann coming across the ice in a Model T from Garden Island gave chase to a coyote. The race between the coyote and car ended on the icy harbor of St. James, but not before the car made several spins and a case of canned bear meat which Kleinheintkz was bringing back to Beaver from his Garden Island home was smashed. The payoff came when one of the men jumped from the car and struck the coyote with an ice spud. This is the latest in Island sport — chasing coyotes in a Model T!

(Feb. 25, 1948)

* * *

After all votes were counted Tuesday, it was discovered there was only one Republican in all of Peaine Township. The question is, who is the lone wolf?

(September 22, 1948)

Thursday, 30 Christian Brothers arrived aboard the North Shore for a retreat at the Brother Domnan Lodge. The Brothers who comprise this group have come from New York, Philadelphia, Baltimore, Texas, Mexico, Canada and Rome, Italy. They will remain until August 20.

(July 7, 1948)

Lucy Norton, aged Indian, was brought back to the Island Friday for funeral services at Holy Cross church and burial on Garden Island. Lucy, 84 years of age, was the daughter of Chief Antoine Peaine. Peaine Township was named after Lucy's grandfather, also an Indian Chief.

(Nov. 10, 1948)

Raymond Lewis crossed the ice to Garden Island Saturday, the first man to cross since the ice made this year. Raymond went across on foot, tapping the ice as he went along as a safety precaution.

(undated)

Nestled among the pines and evergreens about a quarter mile from the village of St. James, Beaver Island, the Brother Domnan Memorial Foundation is located. It is a summer vacation retreat spot for Christian Brothers and stands as a tribute to a devoted Brother who spent his life teaching religion and training altar boys at St. Patrick's school, Chicago. In 1928, the land on which the retreat house stands was donated by a former Beaver Islander, John P. Maloney, now of

154

Chicago. The same year the gift was accepted by the brothers the project on the Island was begun. Financial assistance was given by the alumni of the Christian Brothers' schools.

(Sept. 22, 1949)

At the tavern the other night one of the visitors, a former barber, gazed around and noticed the fringe over the ears and down the backs of so many men. He decided to do something about it. He called for barbering tools and trimmed up about 17 Islanders.

When the job was finished there was enough hair on the floor to stuff a mattress.

(Aug. 6, 1947)

The geologists who were here recently discovered a rare piece of marble on an antique table in the King Strang hotel. The marble, which is dark color, is embedded with a variety of fossils, marine plant life, pieces of shellfish, stems of star fish, small sea animals and coral. They estimated the marble to be a million years old.

(July 2, 1947)

After years of trying, piano owners succeeded in enticing a piano tuner to come to the island. Mark Baird is here to stay until all the "tin-panny" sounds are out of Island pianos.

(undated)

The Beaver Island Board of Commerce and the Game Club will work together to clear the harbor of unsightly sunken boats, piles of old docks and to clear the weeds out.

(Feb. 25, 1948)

A balloon which was released at Lake Forest College, Ill., at the kick-off between Lake Forest College vs. Illinois Wesleyan Oct. 25, landed on Beaver Island at Sucker Point. Carl Kuebler, down at the North end checking the traps set for coyotes, noticed the orange colored balloon caught near the beach.

(Dec. 15, 1948)

'Tis a bit late but this is news: Beaver Island is no more a hot bed of Democrats. For the first time since 1932, the island went Republican. What turned the tide? According to one of the former died-in-the-wool Democrats, "It's those fellers in Washington. It's high time all true Americans wake up to a few facts."

(September 28, 1950)

* * *

Seeing a horse and wagon tied up in the village gave Dick Prendergast and his buddies an urge to take a ride. A deal was made with the owner of the one horse contraption — a bottle of beverage for a ride — so John Ricksgers said, "Hop in." The boys didn't know what they were asking for.

No sooner had they taken the reins in hand and persuaded Old Dobbin to get going than one of the rear wheels broke off and down went the joy riders in a heap. Well, they had a heap of fun and John said he didn't mind, he got his treat and he was satisfied.

(August 3, 1950)

Tattle tale taxi tracks traced Beaver Island's only taxi to the Coast Guard Station after it was discovered missing from in front of the Shamrock Tavern. The culprit found out that crime does not pay. The car's pipe line froze up and he had a long cold walk back to the village.

(February 9, 1949)

From the looks of things around town the morning after Hallowe'en one would think a tornado had hit the Island and put our buildings in freak places. A fish shanty was on top of Archie's Algoma cabin; a Chick Sales cabin stood in front of the Post Office; another was on the deck of the North Shore! Any number of them were turned over; a large row boat was across the side walk in front of the Shillelagh. The old timers say it isn't anything like it used to be. To hear them relate tales of past Hallowe'en tricks this year was mild in comparison.

(November 5, 1947)

Oh "deer" what can the matter be! Mr. and Mrs. Archie LaFreniere returned to their home Friday evening after a birthday celebration at the Shamrock. When they entered their living room, the sight they met stopped them dead in their tracks. On the davenport was Archie's buck, dressed in a nightie and other unmentionables, in sweet repose. The buck, which had been hanging in the yard, was missing for several days. The LaFreniere's feared the ole deer was gone with the wind or somewhere. They hoped by some hook or crook the culprit had retriever instincts and would return with "deerie." Needless to say the pranksters must have had the time of their lives dressing up that big piece of venison for the return trip home. Says Archie, "That was my biggest and fanciest birthday present and I might add the most surprising."

(December 22, 1949)

Governor G. Mennen Williams flew to Beaver Island Friday morning and took the islanders totally by surprise. The trip was the result of a practical joke. Early in the spring, a rumor was circulated the Governor was coming to the Island for the purpose of buying a house. The rumor was the "brain child" of Johnny Andy Gallagher and Elston Pischner. It was one of those dull weeks of the thawing season and the two decided to liven up the Island with some exciting news. A fake telegram was drawn up and the Islanders swallowed the message, hook, line and sinker. Preparations were started to make the governor's visit one he would never forget. Everything was set and the period of waiting became tedious. Finally it was discovered it was all a joke. The next week, the joke was published in the Courier under "It Happened on Beaver Island" and the story got around to the Governor. It was then he decided to visit Beaver Island. Taking off from Mackinac Island in a Widgeon amphibian plane Gov. Williams, with Mr. and Mrs. Francis Stebbins of Lansing and Tony Barnum, pilot from Alpena landed on the beach in the heart of the village about 11:00 a.m.

Archie had a little notice. He dashed home to put on a white shirt and bow tie. As he entered the Shamrock with a tall young fellow, Bert McDonough, seeing the white shirt Archie was sporting, rushed up and greeted him with "Good morning Governor." After a little kidding, Archie introduced the tall smiling gentleman. "Bert," said Arch, "I want you to meet Governor Williams." Well sir, Bert just about swallowed his Adam's apple. Said Mrs. David Wilson after she had been introduced to the Governor, "of all times for me to go to the village with my hair in pin curls." When the "North Shore" docked, the Governor was there like a regular Islander. From off the boat stepped Mrs. Carrie Sonderegger. Seeing Archie, an old sparring partner of hers, she touched him on the shoulder to say, "Aren't you going to greet an old friend?" After the usual how-de-do's Archie said, "Say, I want you to meet the Governor of Michigan." "Now, don't you try to pull any of your Island jokes on me!" Turning to the Governor she said "And young man, the same goes for you." With that Mrs. Sonderegger walked away. But wasn't her face red when later she discovered it was the governor.

(August 10, 1945)

* * *

Well Sir, declared Ole Timer "I never in my born days saw John Ricksgers move so fast as he did last week when Martin's cow got in his basket of groceries." It seems as though John had stopped in the Beachcomber for a little refreshment after shopping and left his horse and wagon tied to a nearby tree. The cows, seeing an opportunity to have a picnic — (it was a wonderful day for a picnic too) — made away with two loaves of bread that were in the back of his wagon and were starting on a package of fish when John rushed up. John saved the fish but there was nary a slice of bread left.

(Oct. 13, 1949)

Young James Gallagher wanted to take a look at the wild horses running at large on High Island, so last Monday, Joe McPhillips flew him over the island to satisfy his curiosity. They had a glimpse of the horses that have lived in the wilds for the past ten years. There were four horses, but only three were seen this trip.

The horses belong to "Little Joe" an Indian, who now lives on Beaver but whose home was on High Island for many years. "Little Joe" makes a couple of trips to the island during the summer to leave salt blocks. Other than that the horses take care of themselves.

(March 16, 1950)

The wild horses on High Island received a bit of publicity lately and the write-up attracted Mr. and Mrs. Leo Willbank of Benton Harbor. They arrived on Beaver Island a week ago enroute to High Island to investigate those horses. They were taken to the Island via Rolland McCann's fish tug with equipment for sleeping out under the stars and cooking on the beach. They returned to Beaver Friday and their experience with wild horses was interesting to say the least. While the Willbanks were away from their camp site, the horses paid it a visit.

They helped themselves to a rain coat and evidently it was very delectable for they practically ate it up. For desert, the horses topped off with a box of brillo pads. In one of the shacks, the Willbanks had put their supply of eggs and other things. These were dumped over and all eggs broken. Exploring around, the Willbanks discovered the missing horse.

There were four living alone and liking it for thirteen years until this winter when only three were sighted. The horse had fallen in a cellar opening and it is believed the animal starved to death. Finding the horse confirmed the belief that some misfortune had happened to one of the foursome. Mr. and Mrs. Willbank enjoyed their camp life on a lonely island with only wild horses for company. They will surely have something to talk about for years to come.

(July 13, 1950)

Theresa Gallagher and the children had a scare Thursday afternoon when James' team returned from the woods without him. On the way home from the lumber camp, James stopped by George Ricksgers, to inquire about Mrs. Ricksgers who was ill. James left his team untied while he went in for a visit. The horses evidently were anxious to get home, so away they went, tearing down King's highway lickety-split. When the team stopped in the Gallagher's yard with James absent, the family was frantic. Thoughts and fears of every description filled the interval between the return of the team and the appearance of James, safely delivered home by George in his jumper.

(Feb. 9, 1949)

The Trail Road became the scene of a runaway when Frank "Big Owen" Gallagher attempted to discourage one of his team from biting at the reins.

Frank, on his way to deliver a bushel of apples, was driving Prince and Randy hitched to a wagon. Prince kept trying to chew at the reins, so Frank decided on a remedy. When he reached Dillingham's the horses were brought to a halt while Frank went a-borrowing lard and a generous sprinkling of red pepper — a sure cure, thought Frank.

Prince, still hankering to chew at the reins, reached over, got one good taste of red pepper and lard and that was enough. In less than no time, all Frank could see was the back end of his wagon bouncing down the road, apples going in all directions. Later on, the horses were found near the Gallagher home caught in a tree, Randy looking rather bewildered, but Prince still burned up.

(November 3, 1948)

"Murder! Kill him!" Went up in cries of all on the dock as he ran off the "North Shore." Hiding wherever possible to escape the wrath of the Islanders present, the intruder eluded capture for one hour. At last Francis Ricksgers cornered and slew the victim and the first rat on Beaver Island in years was thrown in the lake.

(January 5, 1949)

This reporter is in the dog house as far as a certain puppy on the Island is concerned. In reporting the killing of the rat recently, no mention was made of Prince, the rat terrier belonging to the Vernon LaFreniere family. It was Prince who scented the rodent hiding behind the baled hay on the dock that led to the capture — and Prince didn't like it one bit that he was not given his just due.

(Jan. 19, 1949)

One freezing day this past week a Beaver Island family discovered four of their eleven pet chickens had frozen to death. One look at the remaining seven chickens and the family decided something had to be done immediately. The solution was to put them in boxes around the kitchen stove. This was done. The chickens looked so comfortable it was decided to let them remain in the house overnight. Thinking they had their feathered friends safely barricaded for over night, the household retired, chickens and all. Early the next morning, a bang on the floor and a couple of squawks awakened the family. In a flash, everyone dashed to the kitchen and what met their eyes they will never forget — seven big chickens on the kitchen table going after everything they could peck their way through (one was about to choke with a big gob of oleo). Panic is a mild word to describe what followed.

(undated)

* * *

Tuesday evening around suppertime the village was alarmed when a blazing fire broke out in Herman Pischner's barn. The fire was not discovered until it was beyond control.

Everything in the barn was destroyed including farm implements, sacks of feed, and a load of hay. The loss was estimated at around $2500.

(August 17, 1950)

A good many Saturday night baths were postponed due to the Martin's fire, but some of the fire fighters had a sweat-bath working close to a blazing house. Cooperation runs high in a small community when there's trouble. Everyone turns out and is willing to help. That's the way it is on the Island. Jewell Gillespie jumped into one of the windows of the burning house to assist in directing the fire fighters. A trap door below the window was open and Jewell landed in the basement. Jewell said, "I got out much quicker than I went in." Buster Elms' father rushed to the scene with his bucket. A veteran in fighting fires with a bucket brigade, he came prepared.

(undated)

Bruce McDonough, Donald Cole and Donald McDonough went sucker scooping one evening last week. With their tub filled, they set about to smoke their catch. After the usual preparation, the fish were placed in Vesti McDonough's smoke house and this venture not only smoked up the house, but burnt it down, fish included.

(May 12, 1948)

* * *

The romantic part of winter time on Beaver Island is the sleighing time. Through the village and down quiet country roads, the merry jingle of sleigh bells announce there is some-

161

one coming. After a week or so of sleigh bells you can tell who is coming by the sound.

(February 25, 1948)

Skating by firelight is a "must" with the young folks just as soon as the ice makes slick and thick enough. Monday night, a stack of Christmas trees was collected for a big bonfire. One lone shanty was pushed out on the ice during this week. It was Owen Chapman setting up for his winter work — supplying customers with perch. A few fishermen were out fishing without shelter on the days the winds diminished.

(January 19, 1950)

Some of our Beaver Island lads are fine dog trainers. Jimmie Lee Martin and his cousins, Marcus and Billy Martin and Douglas Burke have speedy dog sleds. They go flying up the road standing on their sleds with the greatest of ease.

Bub Burke, the rural delivery mail man has a comfortable arrangement for making the rounds. On his sleigh, he has a small cabin with a stove inside and he is as warm as a bug in a rug as he drives along the country trails. Saturday, Mr. and Mrs. Bob McDonough and family boarded the mail sleigh to move to the village for the next month or two. They will occupy an apartment in the Jewell Gillespie home.

(February 23, 1950)

Beaver Islanders were treated to a gala night Thursday at the "Killarney Klub" with Spalding's all-star cast entertaining a packed house. The Parish Hall, transformed into a nightclub for the occasion, took on the full effects of a swank place to dine and dance. Individual tables for four centered with burning tapers were placed in semi-circle arrangement in front of the stage where the best floor show in town was the main attraction.

(February 23, 1949)

* * *

Regardless of how long one has lived on the Island, the sight of the first spring boat never ceases to be a thrill. Tuesday, the "North Shore," piloted by Mark Cross, made it to the Island sixteen days ahead of the first boat last year. Although the "Mesquite" plowed a path through the harbor ice, the boat

162

could only get within 100 feet of the dock, so Cross unloaded at the point. Friday when the boat arrived there were still plenty of ice chunks in the path and ice two feet deep from path to deck. Unloading was on the ice this trip.

(April 7, 1948)

The netshed at the left is now the "Marine Museum."

NAMES AND PLACES
of
BEAVER ISLAND

There have been about five thousand "Beaver Islanders," people who've spent the most significant part of their life on Beaver Island. Some of their names have occurred in the preceding chapters. Other names are used in island conversation as reference points: " . . . his mother married a cousin of the man who lived with so and so's uncle." Still other names are used to identify places: " . . . down by Jerry Corbett's." "Who's he?" "I don't know, but my father and his father always called that cabin Jerry Corbett's."

This final chapter is an index to some of those names. It can be used to either identify a name or a place; such as it would be a good companion on a tour through the countryside. Undoubtedly it contains errors and omissions. We would be pleased if these were brought to the attention of the Beaver Island Historical Society so that future editions can be corrected.

On the maps, of St. James, houses shown with dashed lines are no longer standing. Those shown with solid lines are, as of 1976.

***

Box A. This area was called Whiskey Point (193) from at least as early as the late 1840's, when about 30 men and three families wintered here. Alva Cable had a trading post here from 1840 (12), which was run by Peter McKinley, supposedly a relative of the President, during the Mormon Era. McKinley added a P.O. to the store. The Battle of Whiskey Point consisted of one cannon ball fired into the water from Troy (301A). The store passed to Dormer, who ran it for 20 years. Then John Day ran it, moving from (105) to (26). His son's wife Myra taught school on the point in a building later moved to (79). He sold the store to James McCann, who lived at (23), a "Kirby Home," and then at (6). Shoppers coming from the country frequently left their wagons at Johnson's (217) and crossed the mouth of the harbor in a small boat.

In 1856 the Harbor Light (9) was erected. Twenty years later the Life Saving Station (10A) was started, with "Tip" Miller as the first chief (see 232). The Life Saving Service

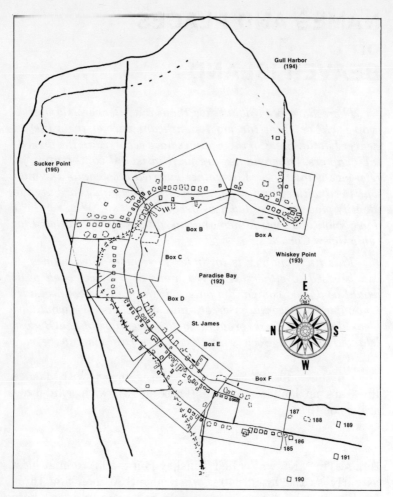

Gull Harbor
(194)

Sucker Point
(195)

Box B

Box A

Box C

Whiskey Point
(193)

Paradise Bay
(192)

Box D

St. James

Box E

Box F

E

N   S

W

187
188
189
186
185
191
190

MAP I

later became the Coast Guard. In 1887 "Big Owen" Gallagher
(22) took over the post and manned it for about forty years,
relying on local fishermen as volunteer help. It's been said
that after he retired he lived to become the oldest retired
"Coast Guard" in the United States. At this time Beaver Island was becoming the leading commercial fishing center in
the Great Lakes, and "Big Owen" and his crew had the bus-
iest station in the country. Next to the light there was a
brick house for the light keepers (8), last occupied by Mr. &
Mrs. Winters before Captain Bennett, who lived at (17), tore
it down. About 1933 a crew barracks (11) was added and a
boathouse (10) was built six years later.

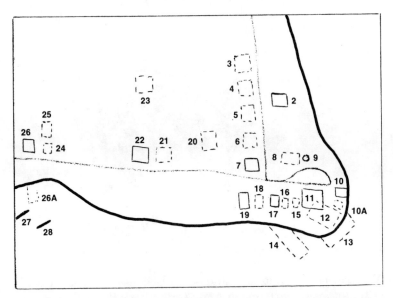

Box A

Although the names and exact locations of most of the early inhabitants of Whiskey Point are apparently lost, a few identifications from the first few decades in this century can be made. We know Daugherty moved into town from the country (261) to a house on the site where Johnny Quinlan later made a house (22) by moving two B.I.L.Co. houses here and putting them together. Later Warren Townsend moved here when he sold High Island to the State, which he'd been using as a grazing ground for cattle. Across the street there was a string of five houses (3-7). "Val," probably a son or grandson of Val McDonough of (316), a night watchman at the B.I.L.Co. (see Box C), lived in either (3) or (4). Tom Douglass lived in (5), and later Tom Bonner (see 281). Joe Left and his son Frank had a house on the corner, where John Andy Gallagher's is now (7), a net shed across the street, and a dock (14). Rouse and Christy Gallagher moved in from Rouse's father's, "Big Phil" (318), to (20), so that Rouse could run the engine in the Lefts' tugs. A fisherman named Pease lived at (21). John Day's home was sold to Andrew Gallagher — "Andy Mary Ellen" —, "Johnny Andy's" father, a fisherman and Coast Guard light ship tender. He built a dock and big net shed (26A) in front of the house. Next to him Everett Cole built a log cabin (24) in 1932; behind the cabin was Johnny Lightning's home (25) — possibly James LeChance's — from before the turn of the century. Johnny Lightning married

167

The 1876 Life Saving Station. Charlie Martin moved it to (10A).

Mrs. Herrick, a midwife, and sold his house to Jock Gibson's (43) brother Robert, one of Andy Gallagher's fishing partners.

John Quinlan, a ship's carpenter and the first Coast Guard here, built a house on the back beach but moved it to (19), next to the site of the island's first saloon (18). John O'-Brien (97) had the next house (17); then Neddeau (16), a fisherman who married Willy Belfy's (48) sister, had a house, which he sold to George Burrel, owner of the "Silver Star." Next there was Captain Herman Ludwig's house, Billy Belfy's stepfather. Captain Ludwig, a boat builder, built the "Edna May" for Belfy, and owned the "Walaska," which sank at (52). Tom Olsen had a house where the C.G. Station stands now. The two sunken ships which were removed from Whiskey Point about 1960 were the "Hattie Fisher" (27), Father Gallagher's schooner, and Lester and Raymond Connaghan's "Seagull" (28).

———— ◆ ————

Box B. This stretch of the shoreline was occupied mainly by fishermen. Gillis Larsen came over from his homestead on Garden Island and built (30). His son "Big Art" had Jerry Palmer (86) build a house next door (29), where Karl Keubler has lived the past 30 years, for him and his bride, Sybil Tilley (84). The house that used to stand on the other side belonged to Joe Smith, who was related to LeBlance (44).

With the coming of the steam tug in the 1880's, which enabled fishermen to reach distant grounds, many of the men who'd been fishing in front of their country homesteads

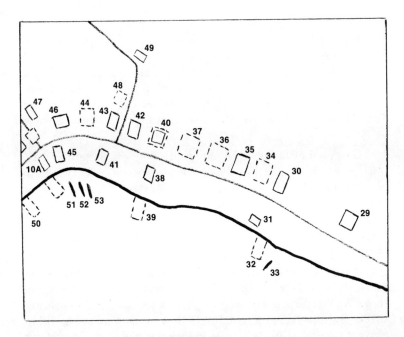

Box B

moved to town, where they could keep a tug in the harbor. The two Martin Brothers, who owned the first steam tug, the "Clara A. Elliott," started this trend. Dan, the elder, lived in the last house built by Charles Tilley (84); it stood at (40). His younger brother James — "Shing" — built one at (36). James married Nellie Johnson (217), and their children moved alongside, Wilbur (named after Wilbur Gill) at (37) and John at (35), in the Barney and Mary Ellen Gallagher home (see 26). Dan's house was built into the back of his net shed; Wilbur's was a net shed moved up from the beach. James' net shed (38) has just been renovated and converted into a Marine Museum. His dock (39) stood in front; the Lakeside Fish Co. stationed a buyer here.

Another of James Martin's sons, Charlie, moved the old Life Saving Station from the point to where it now stands (10A), in front of his dock (next to one which had belonged to the B.I.L.CO). Fifteen years ago three old wrecks could be seen to the south of this dock: the "Walaska" (51), Captain Allers' "X-10-U-8" (52), and the "Elliott" (53).

The "X-10-U-8" was owned and operated by Captain Allers, who arrived on the beach with a load of rotten apples and stayed. After awhile he had 48 grandchildren within a few blocks. The Captain had a house at (41) built by W. C. Wright, but traded homes with his son-in-law Gus Mielke

169

(42), who ran a lumber mill. Cap Allers' son Herman built a house right behind (42), but it burned. Irwin Belfy, who married Leonora Allers, had a home at (48) but it too burned; then he bought the property at (47), which had been a store, first owned and run by Gibbons and then by Nels LaFreniere, before his marriage to Sophia Boyle (151), but had burned. Herman Pischner, another of the Captain's sons-in-law, lived at (49).

George LeBlance, who married Joe Sendenberg's sister (46), lived in his father Paul's log cabin at (44), with his brother Joe. Paul fished with James Martin. When Joe Smith (34) passed away his widow moved here. Jerry Palmer (86) also lived here. Finally Joe "M'Fro," Joe Sendenberg, moved the cabin in back of his own house (46) and made a barn out of it.

William "Jock" Gibson had a house at (43), later owned by Jim "The Bay" O'Donnell. The night Prohibition ended there was quite a party on Beaver Island. Late in the evening three Indians (Frank Norton, "Bow and Arrow," a fiddle player; George High, who was married to a daughter of Harry Lewis (70); and Pete Peter Manitou, Pete's son — old Pete himself was considered almost a genius in many fields) set out for Garden Island in a small boat, but didn't make it. Their bodies were found the next day. A wake for them was held here (43) which rivalled those held for Irishmen.

The fine house at (45) was originally one of the net sheds which ringed the harbor. It was built by Rolland McCann.

Jimmy Gordon rented the dock at (50) from the Beadles Fish Co. of Bay City, which did its Beaver Island Business off a dock at (126). He had a net shed here as well as at (107).

The wrecked schooner at (33) was Charles Cross's "Walhalla," 96 feet long, built in 1867.

Box C, the Beaver Island Lumber Co. On December 27, 1902 three men, George Kitsinger and the two Stevens brothers, organized the Beaver Island Lumber Company with a capital of $75,000. A mill with an initial daily capacity of 30,000 feet of hardwood was built at (75). To bring timber to the mill a narrow gauge railroad (72) was laid, leading from camps along the west side where the company had purchased timber rights from McCrea, to town. Coal for the locomotives was kept at (77), while the locomotives themselves were kept in a roundhouse (71). Behind the roundhouse they built a blacksmith shop (73). Some work was done with horses, which were kept in a large barn (58) with a pumphouse (57) nearby.

The company built two strings of houses for its labor force, which was 75 men in town and another 75 in the country at the start. One row was along Freesoil Avenue (60-70), which was named for the home town of the Stevens brothers. The other row ran for a block to the west, from (80). Part of the house at (79) had been a school on Whiskey Point before John Johnson moved it across the ice; after it was moved and enlarged Bill Stevens lived here. Dr. Branch lived and had an office here when he was the island doctor. It was also McCrea's home at a time when he owned 8500 acres on the island.

Jerry Palmer, a plasterer who was born on South Fox Island and whose father had been on Beaver before the

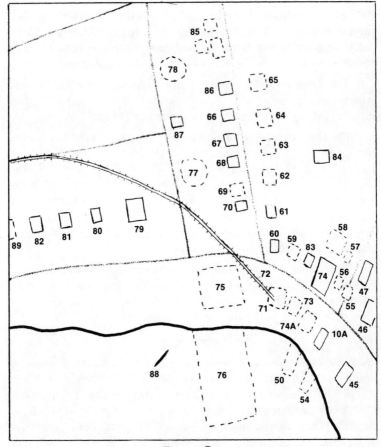

Box C

171

Mormons, lived in (66) and built a house next door (86). To
the north there was a small settlement of Indians, including
the Chenotin brothers, Louie and Bundy. Amelia Bird lived at
(87) before going to Hollywood to seek her fortune; she was in
"Indian Love Call." Across the street Angelline Wabaninkee
lived with one of her husbands, Peter Taylor. "Grandma"
LaFreniere, Nels' mother (150), kept boarders at (67). Harry
Lewis, who lost an arm working in a later mill, lived at (70).
And Joe Floyd and then Maria Gallagher lived at (68). Tom
Hunt, a dapper man who ran the hand car every day of the
year, repairing rails and ties, lived at (60).

The lumber company bought the store Wilbur Gill had
just built (74) and used it as the company store. They put up a
storehouse across the street (74A) for grains, and put in a
softwood dock at (54). They then built a tramway connecting
this dock and other docks to the east with the mill.

At the mill site they built their central dock (76), an immense structure that reached 500 feet into the harbor. Company ships stopped here to load the company lumber. The
sawdust from the mill was taken to (78) and burned.

The house at (56) became the company cook house and
was run by a Piper (282). Next door (55) was the men's dor-
mitory; later this building was moved to (119).

On the other side of the store, next to the lumber company's house at (59), Charles Tilley built himself a home, the
first of many houses he was to build on the island. He and
Noe Stebbins (who lived at (56)) had a mill on the beach before the company came. Tilley homesteaded the land to the
north, building himself another home (84) in order to live on
his homestead. To the east of him a man built a log house and
tried unsuccessfully to raise mink.

The wreck of Alice Coffey's steamer, the "Annabel," once
rested at (88). It was last used by the Martins as a coal barge.
In the early 1960's the rudder was dug out of the sand and
raised, and now is inside the Marine Museum (38).

Box D. The end of a block of lumber company houses is
shown on the right of this box: 89-92. Billy Sheid, a book-
keeper for the company and later purser on the S.S. Kansas,
the only passenger ship on Lake Michigan during WWI, lived
at (92). Hugh Connaghan moved to (91) from (324), and when
it burned moved to (89). The island's first Protestant church
was built at (93) for the B.I.L.Co. employees from Freesoil,

172

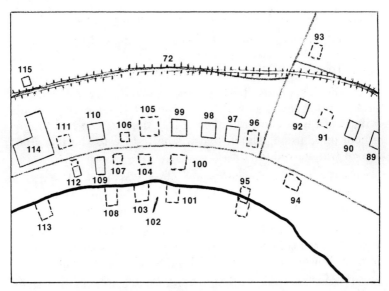

Box D

Michigan. Later Tom Gatliff, one of the island's quickest nicknamers, moved it to (120) and made it into his home.

(96) Biddie Sam's. Biddie Sam, a daughter of Big Dominick (370), married Sam Dunlevy (288), and lived here with her two daughters, Nellie (McCafferty) and Mary Og (O'Donnell, Johnny Og O'Donnell's wife).

(97) Wilfred O'Brien's. Doctor Graham lived here, but when he died Wilfred O'Brien, who had a garage at (94) and lived at (82) before the Cornstalks, married his widow.

(95) Dennis Boyle, Dan Boyle's son, lived above his net shed here, which was partly on his dock. He and his brother Hughie "Malago" boarded at Og's (96). Dennis then married Sam Hunt's widow and moved to (81).

(98) This house first belonged to John Stevens of the Lumber company. Then the James "Bowery" Gallaghers ("Old James") bought it (see 180).

(99) Fred Sendenberg's; Fred was a fisherman who had a dock in front of his house (101). He married Susan Boyle, a sister to W. W. Boyle. Fred also fished in the 'Lake of the Woods' in Canada. He fished pound nets here with Bowery (180).

(100) Val lived here before moving to the Point (3). The Gordons lived here too before it burned around 1912.

(102) The wreck of Dan Boyle's "Mary Ann," an 18 ton

173

schooner built in 1885, rested here for many years.

(103) Yankee Jim (141) owned this dock. Cull and Connaghan's "C&C" tied up here, in front of Yankee Jim's ice shed (104).

(105) This house first belonged to John Day, a fisherman and storekeeper. Later Peter O'Donnell lived here, and his wife, a sister of John Maloney (241), Mamie, who had taught school on Garden Island, had a small store to the left (106) which was moved to (156).

(107) Jimmy Gordon's net shed was here. He was a fisherman who had the "Lily May," the "Meadow Mac" (with his son Robert), and a steam tug named the "Badger."

(108) Charles Pratt (147) and his son Forrest — Captain Pratt — used this dock. Charles married Molly Boyle, a sister of both Bowery Bill's wife and Fred Sendenberg's wife.

(109) This was Anthony Malloy's second butcher shop, run, except for the first year, by his son L. J. Their slaughter house stood to the rear on the side towards Gordon's; the cement slab, with its blood gutter, can still be seen.

(110) Anthony Malloy bought this home from Beaudoin, the cooper; his son Phil Beaudoin earned the caption "quick as a cat" for his actions saving the crew of the "Queen City," wrecked on Hog Island reef with Andy Gallagher and James Martin under Big Owen in the "Elliott."

(111) A Mormon house stood here past the turn of the century.

(112) Mary Bonner (281) married Harry Hardwick (266) after first marrying Lanty McCafferty, who died. Harry was a carpenter, who built Daniels' house (247), while she ran a candy store here. When the road was widened the meat market next door was moved back but the candy store just had its front cut off.

(114) The Beaver Hotel, now the King Strang. This hotel was built in 1901, and opened in 1903, by Captain Mannus J. Bonner (281), at a cost of $10,000. The architect was Father Zugelder. The lumber was boomed from Hog Island and cut at the Stebbins and Tilley Mill. The Captain had a dock across the street (113) in order to offer sailing tours to the other islands in a little sailboat. He also took his guests for jaunts in the country in his buggy. Two horses he bought from Pete Manitou swam back to Garden Island when he let them run in the spring. Initially Bonner planned to build another wing,

174

so that the entrance would be in the center, but business didn't warrant it.

(115) This home was built by Beaudoin for his married daughter; later Paddy Mary Ellen, Andrew Gallagher's brother, lived here. The founder of the Reese Coal Co. hung himself here before the turn of the century. The beach behind this home was known as "The Portage" for a hundred yards in either direction. In the summer Garden Islanders beached their boats here, and in the winter they walked or rode their horses across the ice.

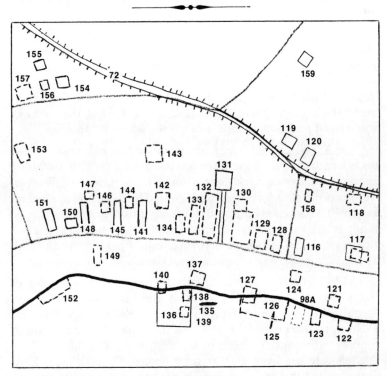

Box E

Box E, Downtown. (116) The building which is now the Post Office was built by Johnny Green as a saloon and leased to the Silver Top Brewery of Milwaukee, under the proprietorship of W. W. Boyle (170). Then it became a tailor shop downstairs and the Ket Gillespie residence upstairs. Later Jack Gordon used the building as a net shed. Finally W. W. Boyle bought it and turned it into a Post Office.

(117) Floyd's. The Floyds, a family which goes back before the Mormon era, had two homes here, one after the other,

a livery stable behind (118), and a net shed (121) and dock (122) in front. Two Floyd brothers, Sam and Frank, were the last to profitably operate the fish-oil business on the island.

(123) Barney Martin, one of James's six brothers, had a dock here. He was the engineer on the "Margaret McCann" and had a boat of his own, the "Silver Moon." A natural rhymer, he used to deliver whiskey to Sand Bay on horseback and make songs out of the news. He lived at (119).

(124) Peter Owen, Owen McCauley's (298) son, built a barber shop and would walk into town every Saturday to give haircuts.

(126) This old dock, built on the wreck of the "Tracy" (125), was called "Beadles" because the Beadles Fish Co. had a buyer on it before 1900. Cundy built a dance hall half on and half off (127). Later the Lakeside Fish Co. and then the Walker Fish Co. had men here. When it was torn down to build the present yacht dock, part of the keel of the "Tracy" was raised and now stands outside the Marine Museum.

(98A) When the yacht dock was enlarged another old dock was torn out to make room, the Booth dock. It had first been owned by the surveyors, Dinsmore, Wright, and Winter, but they'd sold it to Jim and John Dunlevy in 1881. The Booth Fish Co., the largest fish merchant operating on the island, bought it from the Dunlevys and stationed first James Bowery (98) and then his son, "Young James," on it as their agent. Young James bought it from Booth in 1933.

(130) Cundy's Saloon. Cundy Gallagher had a cluster of businesses between the Post Office and the livery stable, the largest of which was his saloon. Johnny Salty ran it for him, although Darkey Mike (273) and Shorty Richardson, a painter who married Hamrock's daughter, also tended bar. Attached to the back was an area enclosed with a solid high fence, called the bull pen, which became the romping ground for those who felt too strongly.

Across the street Cundy had a dance hall, where he held two dances a week. Next to his saloon he had a barber shop (128) and a tailor shop (129); one of these buildings was built around the pilot house of the Milwaukee Belle (396). All of Cundy's buildings were destroyed in a sweeping fire the night of the corn wreck on Gull Island Reef.

(131) The building Stan Floyd uses as a car rental was originally Lou Briggs' Livery Stable. When he died his widow moved to (60), where she lived with Lou Paddy Ruah.

(132) The bowling alley. This long, skinny building was built and run by John Grill, who bought the B.I.L. Co.'s store (74).

(133) The two story building which once stood next to the bowling alley contained different restaurants and shops downstairs, including a bar and Agnes O'Donnell's ice cream parlor (which couldn't compete with Nels'). The upstairs was a residence; Mabel Cull lived here, as did Dominick Big Dominick after retiring from service at Beaver Head Light. Excavations behind the building yielded as many as six very old human skulls each time (Dr. Ruth kept one on his piano), each cracked in precisely the same place.

(134) Big Neil's store. Big Neil was a brother to Big Owen and Big Dominick. He also owned a fish tug, the "Violet," which Ed Martin ran. He lived at (157).

Downtown St. James (134, 141, and 145).

(135) The last resting place of the "Tom Paine," Andy Roddy's 72 foot schooner built in 1871.

(136) The McCann Dock. As the downtown grew and the prominence of the Point declined, James McCann decided to relocate. He gave up his store on the Point and in 1887 built a

177

store (141) in town. Later he added a dock, net shed (137), packing shed (138), ice house (139), and freight shed (140). The store was very modern, having electric lights run by a generator, (electricity didn't come to the island until 1939). The upstairs was built as an opera house and social hall. Father Zugelder held Midnight Mass here, and it served as a jail when George Williams was sheriff. Mike McCann ran the store and dock while his brother John ran the "Margaret McCann," the island's biggest steam tug.

(142) This house was owned by the McCanns, but "Shivery," Johnny McDonough, who worked for McCann on the dock, lived here. In the winter he built jumpers. He owned the "Ella May."

(143) Gibson's. The Gibsons had owned the Mormon Print Shop (now the Museum), running it as a hotel, the Gibson House. Eva Gibson married Jack Dunlevy Jr., and this then became their home.

(144) This was Jack Dunlevy's shoe shop. Many Dunlevys were shoemakers. Later Shivery built his jumpers here.

(145) Yankee Jim's — The Beachcomber. This building was built by Yankee Jim Dunlevy as a grocery store after he returned from Chicago in 1900. Yankee Jim arrived on Beaver Island in 1859. Later he became a shoemaker, but in 1871 he went into general merchandising. Bad breaks and recessions forced him to work for John Day in 1878. In 1888 he moved to Chicago, started his own business, and returned as a successful man. His wife was the daughter of William O'Malley, who operated a store on Beaver Island for a few years starting in 1861. Yankee Jim lived upstairs of his store.

Later James McCann, the grandson of James McCann, bought the building and turned it into the Beachcomber Tavern.

(146) This was Jack Dunlevy's home. Mabel Cull and Ellie McDonald also lived here at other times.

(147) Pratt's. Charles Pratt, Captain Forrest Pratt's father, lived here. This was called "Pratt's Tavern" by some, who claim it burned down when a batch of home brew he was mixing blew up.

(148) The Village Inn. Nels LaFreniere built this at (149) as a dance hall and ice cream parlor; later it was moved to its present site shortly after it was converted to a tavern and its name changed to the Shamrock. This is also the site of Anthony Malloy's first butcher shop, which burned.

(151) Nels' Store. For awhile this had been W. W. Boyles' tavern; a dumbwaiter had supplied the card players upstairs. Then Sophia Boyle's brother "Denemy" (who brought pig thistles to the island) set her up in this store in 1907, which had been built, possibly by Big Neil, on the site of the Johnson Store, which had stood in the Mormon era. After her marriage to Nels LaFreniere they ran it together and lived upstairs. Their granary was in the rear. Then Nels bought the house next door (150), which had belonged to Scopp (347) and, before that, Jimmy "The Priest," who'd managed Father Gallagher's affairs. Nels' dock (152), the site of Strang's assassination, had been one of the two docks at which passing steamers had stopped for cordwood. It had been 200 feet wide, and a tramway had facilitated loading the fuel.

(153) The fanciest house the Irish immigrants found on the island stood on this spot, possibly the McCulloch home. Ellie Roddy McDonald, whose husband died with Cornelius Johnson, lived here before it burned.

(154) Mamie O'Donnell's store was moved here and turned into a home.

(155) Bill Pat McDonough's. Bill, a fine mason who built the Protar Memorial, and many of the first sidewalks, lived in this house and started the foundation in front (156), but died before he could build a house on it.

(157) Johnny Green's. The large home which stood here was Big Neil's home. Then Johnny Green bought it and lived here. When he moved to the country (266) he rented it to Burns. Johnny Green moved the shed at (158) to this site. Sheldon Bassett (328) had a home near here.

(159) Napont's. This home was moved from island to island, being dismantled and erected each time, and each time growing larger, until reaching this site.

Another site of interest is the location of the old two-cell jail, across the street and south of (119). (Now it's standing next to the Print Shop.)

Box F, the edge of town. This was the southeasternmost part of the harbor to be developed because beyond here the shoreline is no longer protected and consequently the water is too shallow.

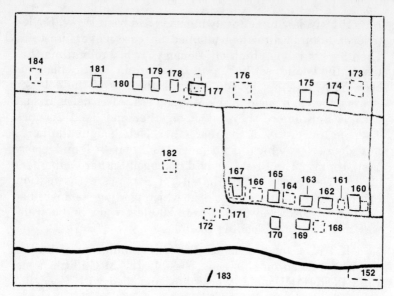

Box F

(160) The Print Shop. This original Mormon building was Strang's print shop, where he published the State's first daily paper north of Grand Rapids. When the Mormons left the Gibsons took over the building and ran it as a hotel for about 45 years. Twenty years ago it was renovated, and made into the county's only museum by the Beaver Island Historical Society.

An addition on the west side, called the Post Office Addition (currently being rebuilt), served many functions. Jimmy Gibson was the Postmaster here. Barber Martin had a chair here, Charles McCann pulled teeth, and Mabel Perron and her husband Tight lived here.

A cooper's shop once stood at (161).

(162) Cundy's. This building, built on a Mormon foundation and possibly once used as a hotel, was Cundy's (130) home. Tessie Connaghan, Cundy's wife's half-sister, bought it and gave it to Jack Cull, her full-sister's husband's son. The upstairs was a hall where labor unions and the Hibernian Lodge met. Dances were held here, and later it became a pool hall.

(163) The Jewell Gillespie home was built at least 95 years ago. It belonged to Tom Gatliff's mother, Annie Conn McCauley, who enjoyed smoking a pipe in the yard; her grandmother, Kitty Gallagher, Big Dominick's mother, lived

to be 103. Jimmy "The Jew" (168) had a store here as well as at (151). It was a three-apartment building: Emma Hunt lived here, there was a dentist's office, and Dr. Palmer had an office here.

Tom Gatliff turned the home across the street (169) into a garage, where he worked on Model T's. But the garage burned. Everett Cole had Ket Gillespie and Peter Johnson build a dance hall, but it fell down. Then the present garage was built. Everett had a store next to it, built on, where he sold fresh vegetables.

(170) W. W. Boyle's. Willy Boyle was the son of William "Whiskey" Boyle, who arrived in 1859 and opened a saloon. Willy was the township clerk, the township treasurer, and commissioner of highways at various times. When he went bald he meticulously trained three hairs to bridge his scalp. He had a story about everything and claimed to have written a history of early times on the island; he may have, but it was never found. He ran the saloon at (116), bought the building, made it the Post Office, and became the postmaster.

Willy Boyle had a pool room across the street (164), which had been one of his father's saloons. Sophia (151) had her first store here in 1905. Then Willy sold it to his nephew, Bowery's son Willy John Gallagher, who lived next door (165) with his wife Mary Duffy, the last Irish Immigrant to Beaver Island. She had worked for Nels at (151) before her marriage.

(166) The building which once stood here was a boarding house. The Sisters stayed here when they arrived. Mary (Greene) Gallagher, Big Dominick's wife, walked into town on a Saturday from Green's Bay to go to church, but had a baby here instead.

(167) When Black John Bonner (281) arrived in 1856 he stayed in a Mormon house here for a year. Then, when he moved to the country, he took it down and brought the logs with him, using them for his barn. Around 1908 the Parish Hall was built here.

(168) Jimmy "The Jew" Gallagher's. Jimmy and his wife, Big Phil's daughter Bridget, moved here from the country (303). Jimmy was a brother to Phillipine (287) and to Shoemaker's wife. He had a little store across the street and also owned two sailboats, the "Up And Coming" and the "Smart Man." In 1910 he went back to Ireland for a week, singing the praises of America's Emerald Isle and bragging that he had his own motor boat. But they refused to believe

that an individual could be so well off as to own a boat he didn't have to row.

(171) Frank Miller built a house here, supposedly buying the continuation of the road from the township supervisor. He had a blacksmith shop here, a small mill, and he also fished pound nets. It's claimed he invented the adjustable hospital bed but neglected to obtain the patent.

(172) Big Mary's house was here. She was old Vesty's (259) sister, who married Pat Carmody in 1862. Pat lived at Cable's Bay in Don McCarty's home before the Mormons arrived. After they married they homesteaded below Hannigan's Road. He and two of the Martin brothers started in for lunch from setting pound nets, and their small boat overturned and all three drowned. The Big Mary moved to town, where she midwifed. When a sailboat needed to be dragged ashore she was ready to help, and could do the work of two men.

(173) Charlie Cross's home was here until it burned. Joe Burke's barn was across the street.

(174) The house that Jack built. Jack Boyle, Turner's adopted son, built this house before moving to Chicago and building many more houses and other buildings there.

(175) This house was built by Scopp, who married Paddy's Hela's daughter Julia (284).

(176) Owen Conn's. Owen Conn McCauley, keeper of the Squaw Island Light, lived here. Music lessons were given here as an extension of the school's program.

(177) The old McKinley School, built in 1901 for $2800, stood slightly behind (overlapping) the present high school. When Beaver Island was part of Manitou County the County courthouse and jail stood to the left. It was used as the school for a year or two while the McKinley school was being built. The high school was in the current kitchen of the Parish Hall (167) then.

(178) Willy the Woodchopper's. This was Charlie Roddy's house. Charlie was one of Captain Roddy's (290) 14 children. He married Margaret O'Donnell, and the house passed to Willy Danny Barney O'Donnell.

(179) Tim Roddy's. Tim was Charlie's brother. He was first mate on the "S. S. Kansas"; there was always a dance the night she was due to arrive, and it went on until she appeared. Then everyone went on board to see a silent movie.

(180) Bowery's. Bowery, William John Gallagher, the father of Willy John (165), was Tight's brother (338). Bowery had worked as a messenger boy on Wall Street, arriving on Beaver Island with his family in 1871 at age 16. In 1872 he began fishing and in 1881 he went into business for himself. He was chairman of the Charlevoix County Board of Supervisors and the longtime supervisor of St. James Township. He married Margaret Boyle, W. W.'s sister, and commanded Tent No. 834 of the Knights of the Modern Macabees.

(181) Tom Bonner's. Tom, one of Black John's sons (281), worked at the Beaver Head Light. He married Cornelius Gallagher's daughter Ellen (296).

(182) The store for the Northwest Trading Company, possibly the first building on the island, was built here in the 1830's. Later Mrs. Ganya and then Uri Winnie lived here.

(183) The wreck of the "Lily Chambers."

(184) Mike McCann's. This large house, one of the finest in town, was struck by lightning and burned.

Map I, conclusion

May I; 1) This abandoned farm was built by Charles R. Wright (related to Isaac Wright, the Garden Island schoolteacher who wrote the ballad, "The Beaver Island Girls," in the 1870's). C. R. Wright was the partner of John Day. It's now called "Bissell's," after Clarence Bissell (who married George Steven's sister (240), who lived here in the 1920's.

(185) Charles McCann, the dentist, had a small house here; it was moved to Peter Johnson's a few years ago.

(186) John McCann's. Charles Tilley built this grand house, as well as both McCann's and Yankee Jim's stores, Anthony Malloy's meat market, and many other buildings. John, an expert ship builder, did the inside trim himself.

(187) Dr. Edgar Ruth's. This was the first summer cottage built in St. James. Once when Dr. Ruth had a broken leg Owen Conn's wife needed him on Squaw Island. He said he'd go if they carried him up the steps of the lighthouse. It was no easy feat, since he weighed over 200 pounds, but they did and he delivered her baby. This cottage burned when a stone fell out of the back of the fireplace in 1959.

(188) This is the site of Strang's home, which was burned.

(189) Miss Graselli's. This house was also built by Charles Tilley. Her family had made money when America was forced to make its own dyes during W.W.I. She and the Ruths came summers with their colored servants.

On the road between (188) and (189) is Bundy's; he brought the deer to Beaver Island.

(190) The Christian Brothers Retreat House, the Brother Domnan Lodge was built by Joe Burke. Although this is now used as a boys' camp, for 40 years it provided a summer place for prayer, meditation, and study for Christian Brothers from all over the world.

(191) O'Leary's Tree House. The O'Leary's lived at Gordon's (242); Mrs. O'Leary was Gordon's sister; they kept horses for the McCanns.

———————◆•◆———————

Map II, Points, Bays, Lakes, and Roads.

(a) Points and Bays

(192) Paradise Bay. This was the name of St. James' harbor (Beaver Harbor) at or before the time of Strang.

(193) Whiskey Point. The site of Alva Cable's trading post, and Post Office, later run by Peter McKinley.

(194) Gull Harbor. A hatching ground for suckers. When Sam Floyd scooped out gravel to sell at 25 cents a load he was following a long-established custom; because the limestone outcrops here, this marl gravel makes excellent road covering.

(195) Sucker Point. A nesting ground for gulls.

(196) Indian Point. A group of Indians lived here in the early 1800's; an Indian cornfield is indicated to the east on an old map. The legendary Forty Thieves allegedly stashed their booty here. Mrs. Williams calls this Rocky Mountain Point in her book, "Child of the Sea."

(197) Cross's Landing. Charles Cross came here to fish.

(198) Donegal Bay. The Mormons had called this Mt. Pisgah Bay.

(199) McCauley's Point. After the pound-netter Owen McCauley (41), who bought this land from James Dunlevy in 1882; also called "Barney's Landing."

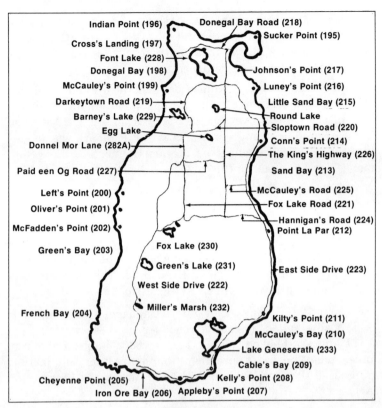

MAP II

(200) Left's Point. Joe Left came from Oconto, Wisconsin to fish here. This is where Captain Hill's body floated ashore.

(201) Oliver's Point. Named after John Joseph Oliver, the son of the founder of the Oliver Tool Co. He was a Scotch fisherman who moved over from High Island and married a Chenotin from Garden Island.

(202) McFadden's Point. McFadden had a fishing cabin here, and O'Brien, an Anglicization of DeBriae, lived here to mark the northern end of the Green's Bay settlement.

(203) Green's Bay. After "White Dan" and "Red Dan" Greene, cousins who settled here in the 1860's to fish. Twelve families, including Dan Boyle (264), who homesteaded the southern portion in 1863, lived here until the 1876 move to the Greenetown area (e.g. 334).

(204) French Bay. It's suspected the French Voyageurs established a base here briefly in the early 17th century. Old histories claim their relics were found here in abundance.

185

(205) Cheyenne Point. One legend is that the fish gave out and the pound netters living here went out to Cheyenne, Wyoming to work on the transcontinental railroad, the point being named when they returned to it.

(206) Iron Ore Bay. Named because the ore in the bottom of a wrecked freighter, the "Betsy Smith" (400), and 100 tons of ore thrown over by the "Barbarian" in 1882, turned the water red in strong southwest winds. Tan bark was loaded here onto such schooners as Manus Bonner's "Rouse Simmons," later famous as the missing "Christmas Tree Ship."

(207) Appleby's Point. After Captain Gilbert Appleby, the third keeper of Beaver Head Light. His apple trees can still be seen.

(208) Kelly's Point. After John Kelly, who arrived in 1863. Kelly was an early settler who bought land and lived here. Pfiefer might also have lived here. The south arm of Lake Geneserath was called "Kelly's Arm" in the 1890's.

(209) Cable's Bay. After James Cable, who with his uncle Alva Cable sailed a trading schooner in these waters from 1837 and established a post (12) in 1838. James started an independent business here about four years later, and in 1848 purchased the land (96 acres for $120) under the Preemption Act.

(210) McCauley's Bay. After "Black Pete" McCauley (81), a fisherman who first worked for Waggley and then, in 1879, bought this land.

(211) Kilty's Point. Patrick Kilty was in the first surveying party (1845). His daughters taught at St. Ignatius School (373) and his son Peter, a Great Lakes skipper for many years, captained the Pere Marquette #18 when she sank in Lake Michigan in 1910.

(212) Point LaPar. One of the very few island names whose meaning is completely unknown.

(213) Sand Bay. Also called Big Sand Bay and, erroneously, Sandy or Big Sandy Bay. In the mid 1800's there were 20 small cabins here, possibly built by summer vacationers brought by schooner.

(214) Conn's Point. Named after Connell McCauley (312), who kept his fishing boat in a sheltered pond just below the point. He built sailboats here which were used to carry Island wheat to Elk Rapids. A wrecked piano which can be seen on the beach here is said to have been one of a load of

pianos washed off the deck of a passing freight boat during a storm.

(215) Little Sand Bay. The Bennett brothers bought land here in 1848.

(216) Luney's Point. Patrick Luney settled here about 1846 after his nephew, Patrick Kilty (211) surveyed the island the year before. In 1852 he sold his farm under duress to King Strang for $40. He became the second keeper of Beaver Head Light (398).

(217) Johnson's Point. Here John Johnson, a Mounted Policeman in Canada before coming to Beaver Island, and his son Peter built the "Nellie Johnson" in 1894. They brought oak logs from Canada for this trading schooner, with which they replaced the "Rough and Ready." This area was also called "Little Canada."

(b) Road Names

(218) Donegal Bay Road. Built on the old railroad bed, paralleling an old Mormon road.

(219) Darkeytown Road. Named after Michael F. — "Darkey Mike" — O'Donnell (272).

(220) Sloptown Road. Possibly named after a road in Ireland. One legend claims the people living here raised pigs; another has it that this was once called "Slaptown" because the gang of boys living on this road were bigger than the boys on Darkeytown or in Greenetown and could — and did — slap them around.

(221) The Fox Lake Road. This was the farthest south any early west side road ran.

(222) The West Side Road. A narrow ridge of gravel, which had been a beach when Lake Michigan was much higher, served as a firm foundation for both the narrow gauge railroad and the road. The first section built was between Camps Five and Four, and was called the "Wheel Road" (the rest was still the railroad).

(223) East Side Road. The original road, called the "Trail Road," was a wagon trail from the Four Corners (west of 305) to Mike Boyle's Beach (80). In 1937 the East Side Road was started as a W.P.A. project. Eventually it included the section south of Hannigan's Road (224), which had been so sandy and had exhausted so many horses that a group of farmers once hauled load after load of clay from north of Tom McCauley's Hill (327) to give it a surface.

187

(224) Hannigan's Road. Named after "Tom Hannigan Boyle."

(225) McCauley's Road. After Tom McCauley. Originally this road stopped at the top of the hill; to proceed south along the east side, one had to open one gate, drive through Tom McCauley's field, open another gate, and drive through Vesty's Field.

(226) King's Highway. After the Mormon leader James Jesse Strang, who industriously developed the island. This road initially had three corduroys, which were felled trees placed side by side to bridge swampy areas. An unfinished extension (351) was cut for use in reaching Lake Geneserath in the winter, when the swamps froze. The entire stretch was called the Enoch Road on Greig's 1852 map.

(227) Paid een Og's Road. The Gaelic, (Pāj´een ôg), for "Young Pat" Boyle.

(c) Lake Names

(228) Font Lake. Named by the Mormons and used for baptisms. A chain of ponds on the southwest side possibly was turned into a canal in spring to lure spawning fish into the lake. Another project begun was the draining of the shallow lake to create a field for wild rice; the ditch was started on the north side.

(229) Barney's Lake. After Barney O'Donnell, whose home is still standing overlooking the lake.

(230) Fox Lake. F. Fox and R. Fox, a carpenter, occupied this land in 1852.

(231) Green's Lake. After "Big Neil" Greene, a cousin to the Dan Greenes, who came over first (1860).

(232) Miller's Marsh. After Harrison "Tip" Miller, a Mormon who stayed after the exodus. He became the fourth keeper of the Head Light and the first chief of the life-saving station at St. James.

(233) Lake Geneserath. It was also called the "Little Lake." The Mormons named it Geneserath, which is the New Testament name for the Sea of Galilee (which it was called on the 1852 map).

Map III

(234) Lost Indian Burial Ground. The Christian Indians converted by Bishop Baraga were buried here; all trace of the graves has been gone for fifty years.

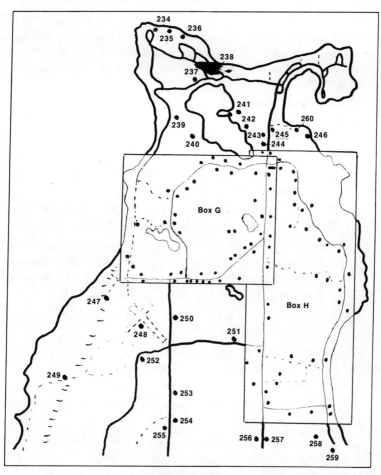

MAP III

(235) The First Catholic Church. In 1832 Bishop Baraga built a crude structure to serve as a church. The following year a more substantial structure was started, but it's unknown if it was ever finished.

(236) Pagetown. A store and a few houses were here; the store saved boats from having to round the point.

(237) Cisco's. Cisco, the B.I.L. Co.'s sawyer, had a house here which burned. Since it was close to the track, the entire

mill crew came out in the train with materials and tools on board. Swarming thicker than mosquitoes, they rebuilt it in one day. H.G. (House Gone).

(238) Fontville. A Mormon settlement, including Jeptha's Castle, the largest house on the island at that time. H.G.

(239) Mt. Pisgah. A dead dune with a view of the Mackinaw Bridge. Named by the Mormons. The cave the Mormons were accused of using to counterfeit money in was supposed to be here.

(240) Steve's Place. Old George Stevens, a cooper from Yorktown, N.Y., built this house, now called "Hidden Valley," or the "Valley Farm."

(241) Maloney's Point and Farm. Pat Maloney came from the mountains of Donegal. His son John taught at (284) and at (79) with Mary Gibson.

(242) Charles Gordon's. This family came from Glouchester. They all returned except Jack, who built himself a small house in town and did some fishing. H.G.

(243) Dennis Cull's. Dennis, a first cousin to Dan Boyles, was Mike Cull's (162) father. The house had the same plan as Bowery's (180). H.G.

(244) Enoch-Enoch Hill. Enoch was another Mormon settlement. H.G.

(245) Frank O'Donnell's. Apparently unrelated to the other O'Donnells, this one fought in the Spanish war. Later Pat Malloy lived here.

(246) John King's. John fished for McCann and lived here. H.G.

(247) Daniels'. Mr. Daniels traded $25 and a saddle for this land, never having heard of Beaver Island. He came here, liked it, and built this house. His Maxwell was one of the island's first cars.

(248) Airport. In the early 1940's the stumps were pulled from the "Pinery," a field of immense pine stumps. In 1943 $5000 was raised to level the landing strip; Tom Walsh landed in the first plane on it.

(249) Peshawbes' Town. A settlement of Indians from Grand Traverse Bay. A few of them worked for the B.I.L.Co. but most of them fished. The women made baskets and picked

vegetables at many farms. The more than ten houses that were once here are now gone.

(250) P. O'Grady's.

(251) Paid een Og Boyle's. Pat was Anthony O'Donnell's (331) wife's brother. Returning from an outing in town one winter evening, he took a short cut — he was staying with his daughter Hannah Johnson (217) — and walked off the end of the dock. His body was churned up by the paddle wheel of a boat the following spring.

(252) Camp #1. B.I.L.Co. camp, on the main railroad line.

(253) Charley Perron's. Charley was Middie's (361A) brother; his wife tended Father Jewell and his daughter married Tight (338).

(254) Cundy Gallagher's. Cundy was Cornelius's brother; he married "Big Rosie," who went down on the "Vernon." When Cundy moved away sheep took over the house and frequently could be seen peeking out the upstairs windows. This area was called the "Black Hills." H.G.

(255) Shamus Gallagher's. Shamy married John Gillespie's (286) sister in Ireland. Shamy's brothers were Paddy Mor (321) and Shawn, who married Ellen Gillespie, Mary's sister. H.G.

(256) Bill Ricksger's. Bill worked at (378) before homesteading here and building this home. He married Bridget Gallagher (275).

(257) Joe O'Donnell's. Anthony's son (331); he had a farm here.

(258) John Gill's. Connaghan owned this property, but John Gill had a mill and a little store here before moving to St. James. H.G.

(259) Vesty McDonough's. Sylvester McDonough, one of the first settlers, built this home here on "Vesty's River" — now called the "Jordan River" — in a week. He bought a boat and nets and started fishing right away, but also managed to operate a large farm. When Vesty died Vesty Vesty took over the farm. One year his potato crop was so good that he bought a Whippet car out of the profits from selling them on the mainland. H.G.

Vesty's home (259).

(260) Troy. A Mormon settlement, perhaps named after the "Troy," a ship which brought Mormons to the island. A Mormon corduroy ran south, under what became the Coast Guard telephone line (removed in 1974).

Box G., Sloptown — Darkeytown

(261) Daugherty's. The original location for this house was farther back. Daugherty returned to Ireland but his daughter Belle married Daniel O'Donnell, living first at Barney's Lake before coming back to her father's house.

(262) Frank O'Donnell's. Frank Danny Barney built this house for his bride, Nellie McDonough.

(263) Owen Boyle's. "Ropa" Boyle, a farmer who came in 1868, married Ellen, another daughter of Big Biddy.

(264) Dan Boyle's. Dan built this after living across the road. He came in 1857. He farmed here and also had a schooner. He charged $7 for a quarter of beef. He was raised with Pete McCauley.

(264 A) Dan Boyle built a house and homesteaded this land. The house burned.

(265) Nackerman's. The first Nackerman house, an old log cabin, is no longer standing. They moved here from (98), buying the land from Jerry Corbett. The house was built in 1902.

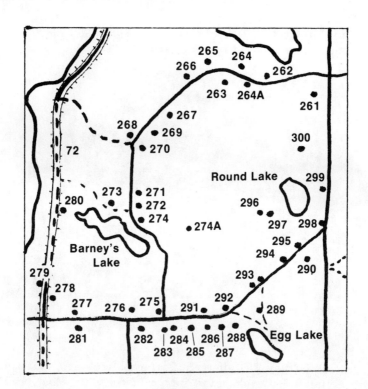

Box G

(266) Hardwick's. Old Henry Hardwick raised bees and took a Mormon, Mary Campbell (306) as his first wife before marrying a half-sister to the Gordons (242). Johnny Green later lived here. Behind the house is "Swenny's Swamp," possibly named for the Sweneys (1860's) with whom Dennis Cull first boarded.

(267) John O'Donnell's. "Johnny Strack" had a log house here bought by John Johnson's (217) widow Hannah and her sons Peter and Cornelius. At the age of 22 Cornelius was shot through the forehead on Mackinac Island and his boat was set adrift, washing ashore on Green Island. Peter built a square two-story house. Johnny Roen, who later became a millionaire, lived with Peter Johnson.

(268) John O'Donnell's. Johnny Barney married Big Emma, another daughter of Big Biddy; two men married to sisters lived across from each other. Barney McCafferty moved here when his house burned. James McCafferty, Barney McCafferty's son, built the house north of this one. Francis Mooney married Jim's widow and moved in here.

193

(269) Jerry Corbett's. Jeremiah, a bachelor from Cork who built this log house, was a carpenter.

Jerry Corbett's (269).

(270) Barney McCafferty's. Barney, Mike's son, married Grace Big Biddy and moved into a house which had been Shamy Gallagher's and "Big John" Bonner's. H.G.

(271) Bonner's. "Big John," no relation to "Red John" or John B, (Black John), retired here in his sixties after working in Pennsylvania's mines. He had a wooden leg and was called "Step-and-a-half." H.G.

(272) "Darkey Mike" O'Donnell's. In another match made by Old Billie, Darkey Mike, a bartender for Cundy (130), married Nanjog, who'd been married before to a Gallagher.

(273) Darkey Mike's Shanty. Darkey Mike stayed here during the time Father Pascal, a strict priest, was on the island. At one time four or five houses were here, hence the name "Darkey's Town."

(274) Barney O'Donnell's. Barney, an elder brother of Darkey Mike and Mary McCafferty, was a farmer who arrived in 1860. He lived here with his brother James.

(274A) Tom H. Boyle's. Tom R, Tom H's son, was Don Burke's uncle. Tom H, the father of Hugh Boyle of Sand Bay, died in 1870. In Ireland he married Mary (O'Donnell) McCauley. Pete Hugh, his son by a previous marriage, married Hannah Beag, McCauley's mother. H.G.

(275) Francis Gallagher's. Francis bought this land in 1862, married Grace Rogers, his second wife, who bore him James F, who married Annie Early. H.G.

(276) Dan "Turner" Boyle. Dan was Paddy Hela's son. He was the second husband of Maggie 'Big Dominick,' who was Johnny Quinlan's mother (19). H.G.

(277) Connoly's-Protar's. Connoly, "Black John" Bonner's wife's step-father, sold this cabin to Fedor Protar, an Estonian actor, editor, and self-educated doctor who retired here in 1893 and for 32 years doctored the islanders without charge.

Connoly's — Protar's (277).

(278) Engineer's Grave. Ed Chase died when his engine overturned in 1913; Dr. Protar erected this cross.

(279) Protar's Tomb. This memorial was built by Bill Pat McDonough, after Paul Kersch raised the money here on the island.

(280) Kuebler's Trail. The last intact stretch of the railroad grade was turned into a walking trail, including the engineer's spring, by Karl Kuebler, the conservation officer, beginning in 1969.

(281) John B. Bonner. Captain Bonner, one of the first post-Mormon settlers, he was a fisherman, farmer, and boat builder. His house, built of Mormon logs, stands just west of the Bonner Centennial farm, east of Bonner's Bluff. He shipped tan bark and lumber to Chicago in his boat, the "Sophia Bonner," of which he was the master for 21 years. He also owned the "Hattie Fisher," the "Rutland," and the "Rouse Simmons." He came from Gull Island the day he heard Strang was shot. He built a dock on the beach in 1856 or '57.

(282) Piper's Corners. "Piper," Dan "Don Mor" Gallagher, Bryan's (378) brother, was called Piper; he worked for the B.I.L.Co. The road running south from this corner is called "Donnel Mor's Lane." H.G.

(283) Kennely's. James Kennely came here in 1871 with his wife, Catherine Cull, and father Pat; he was a wagonmaker and farmer. H.G.

(284) An 1880's School. Johnny Maloney taught in the log cabin here, walking from his home (238) and back for 50 cents a day. H.G.

Paddy Hela's (Harlem's), (285).

(285) Paddy Boyle. "Paddy Hela" ("Hela" is Gaelic for Julia, his mother) was one of many Paddy Boyles on the island. He was a sailor and farmer from Aran Mor who went blind. His sister married Big Phil (318) and his daughter Sophia married Nels LeFreniere (151). "Harlem," Phillip C. Gallagher's son, bought this farm (which is now known as "Harlem's") from Denemy, one of Paddy's sons.

(286) John Gillespie's. John homesteaded this land. One of his daughters married John Connoly (180) and his son Ket, who cooked on the "Ryan," married Mary Anne "Paddy Hela" (285).

John Gillespie's (286).

(287) Phillip C. Gallagher's. "Phillipine" was the nephew of Big Owen's (309) who paid his passage here. He began buying this land in 1882. He married Mary Gillespie, his neighbor's daughter, and had fifteen children. He was township supervisor and sheriff of Manitou County. H.G.

(288) Dunlevy's. Captain Daniel Dunlevy moved his family to this farm in 1859. When he died his son Francis "Sam" maintained it. Dan's daughter Mary married John Gillespie (286); the other sons, John and James, started cobbling in St. James. James became "Yankee Jim" in 1900. H.G.

(289) Harlem's. This farm in front of Egg Lake was run by one of Phillip C. Gallagher's sons, Charlie, a mail carrier on the ice who also worked for the B.I.L.Co. and was known as a great storyteller. The structure at the lake was the barn. H.G.

(290) Captain Roddy's. Andy Roddy was a singer whose voice rivalled John McCormack's; he was so good he "never had to buy no drinks for himself," and he "could sing every damn song there ever was." The nephew of "Black John" Bonner, he was the master of several lumber hookers. He made this his home by 1863. H.G.

(291) "Labbly" O'Donnell's. Mike Cull's grandparents; after many years of blindness he died and she moved in with their son Frank; then Conn O'Donnell, Shamus's father (Shamus and Pat Boyle raced their sulkies up and down the road) took over this house. Conn's sister married Carne and lived here.

(292) Agnes Scott's. Agnes was a widowed sister of John Gillespie. Francis Mooney and Jim Mooney, who wed Hannah McCafferty in one of "Old Billie's" (362) best matches, lived here.

(293) Mike McCafferty's. Mike married Mary, the sister of Barney O'Donnell (274). When their house burned down Dr. Protar gave them hewn timbers to build the barn/house which still stands. When Mike died Mary lived upstairs and her son Francis, who married a Dunlevy, lived downstairs. Pete the Swede, a sailor, built the log cabin here in the 1930's.

(294) Red Hughie's. "Red Hugh" Boyle was "Big Biddy's" son. He had the best garden on the island but no one ever saw him work it. Apparently on his death bed, he was revived by a good slug of whiskey and lived for another six years. H.G.

(295) Bridget McGowan's. James McCauley, one of the five McCauley brothers, also lived here; his family was plagued with t.b. H.G.

(296) Cornelius Gallagher's. Cornelius, from Burton Port (on the mainland across from Aran Mor), worked in the Pennsylvania mines before coming here. "Salty" (342) was his son, and so was Cundy, who had several businesses in town (130). Cornelius's Swamp starts at the northwest corner of his field.

(297) Adolph Ostenbury's. This Swedish fisherman lived just east of Cornelius. He married Celia Gallagher, who'd previously been married to "Big John" Bonner's son John. H.G.

(298) Owen McCauley's. Another of the five McCauley brothers. His son Peter tore down the old log house and built the one now standing. The Mormon town of Siloum was here.

(299) John McCauley's. John was Rae Gilden's father and one of the five McCauley brothers from Aran Mor. He married "Katcheline" Og, who was famous for her beautiful lilting; when the fiddlers tired she took over, and her wordless music was as good to dance to as any fiddler's.

(300) Gilden's. Nels Gilden, a carpenter from Sweden, arrived between 1893 and 1894. His daughter-in-law Rae (McCauley) Gilden cooked at the Beaver Hotel when Manus Bonner owned it, and finally bought it from him and ran it herself. John Dazzler took over this farm and had a dairy here. Then John Ricksgers bought it.

Box H, King's Highway — East Side

(301) O'Donnell's. "Mike Mahane" was one of the Trail Road O'Donnells (315). He married David McCormack's (320) daughter Mary, who later moved to New York with the children. Willie Ricksgers later had a blacksmith ship here after (256).

(302) William Gallagher's. "Old Billie," an uncle of "Bowery" from Tyrone, had a saloon, "The Wayside Inn," here; he's remembered for the many fine matches he made.

(303) Connaghan's. This family was unrelated to the Sand Bay Connaghans. Later Bridgey the Jew, Jimmy the Jew's wife, lived here, and Frank McCauley. H.G.

(304) The first landing field. Bud Hammond, who had the first mail contract to the Island, landed in this and other nearby fields when delivering mail and flying out fish and freight in the 1920's and 30's. It belonged to Ricksger's, whose house was south of here.

(305) Church Hill. This is the original location of Holy Cross Church. The Catholic cemetery, the Rectory (now the Circle M. Supper club), with its 16 inch thick stone walls, and the weather tower were located here.

(306) Campbell's Hill — Campbell's. The Campbells were a Mormon family who were allowed to stay after 1856. Later "Paddy Bacah" McCauley lived in their house; one of the five McCauley brothers, he had three sons all of whom were Captains on the Great Lakes. H.G.

(307) Joseph Schmidt built a new house here about 1900.

(308) James McCauley's. Jim was another of the five McCauley brothers. This was a Mormon house surrounded by

cherry trees An epidemic of consumption wiped out the family. H.G.

(309) "Big Owen" Gallagher's — Welke Airport. It's said that, before Big Owen, the Mormons had built the house around the cabin of a ship they'd found wrecked at Sand Bay. Big Owen was appointed second keeper of the life saving sta-

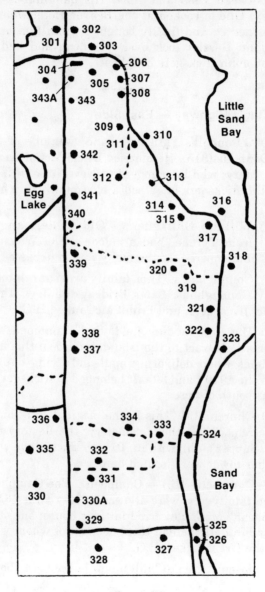

Box H

tion in 1887 when it was the most active station in Lake Michigan. Bill Welke began an airport here in 1966.

(310) John Early's. This originally belonged to Joe Warner, a German soldier. The childless Warners willed their house to John — "Shawn" — Early from Runafasta and his wife Margaret (Sharkey); their son Paddy married a "Daniel Peter" (314) daughter.

(311) Tom McDonough's — The Golf Course. "Big Tom," Mary, who married Owen Conn, and Ann, who married "Red Hughie" Boyle, were Val's children, as well as Jim, Kerry and Francis.

(312) Conn McCauley's. The eldest of the five McCauley brothers.

(313) Phil Malloy's. Phil was the son of Pat Malloy by his first wife; he soon moved to Chicago. H.G.

(314) Dan Gallagher's. "Daniel Peter," ("Daniel Peter" in Gaelic is pronounced 'Donnel Father') was the father of Mary Early and Doney. Dan came over in the large 1884 party.

(315) John "Mahane" O'Donnell's. John, the father of Joe and "Mike Mahane," had dances here because he had the only hardwood floor. Possibly the "Johnson House" which Mrs. Williams said was the largest house on the island in Mormon times, was moved here. H.G.

(316) "Val" McDonough's. "Val," Tom, was a cousin of Vesty (259); he was "Big Tom's" father.

Val and his son Kerry's (316).

(317) Dominick "Paddy Mor" Gallagher's. Son of Paddy Mor (321). H.G.

(318) "Big Phil" Gallagher's. Big Phil was the first Irishman to bring a team of oxen for his farm; his half-brother-in-law Dan Boyle brought the second. "Rouse" was his son.

Big Phil's barn (318).

(319) Dan Gallagher's. Dan was "Doney," who went back to Ireland to find a wife.

(320) McCormack's. David McCormack lived here in a tiny house before 1900. One daughter married "Gebo" and another married "Mike Mahane."

(321) Paddy Mor's. Paddy Mor was a relative of Piper (287). This was his first house. Later it was James Vesty's, a carpenter who married a daughter of Big Phil. H.G.

(322) Hugh "Gebo" Boyle's. James Vesty built this for his family but when first his son and then his wife died he lost interest in moving in, and soon died himself.

(323) Mike Boyle's — Mike Boyle's Beach. Mike, who came to the island by way of Pennsylvania, was a cousin of Dan Boyle (266). His daughter married Buffalo Malloy. H.G.

(324) Hugh Connaghan's. Hugh had a store here in the 1880's. H.G.

(325) Dan McCauley's Dan wasn't closely related to the five McCauley brothers. Dan was a farmer who'd been a

202

schoolteacher in Ireland. He subscribed to an Irish paper and gave readings from it to his neighbors once a week. In Canada he sent for his two nephews, Pete (382), and Dan Boyle (264 A), before continuing to Beaver Island (1857).

(326) Pat Malloy's. A fisherman who was unrelated to the other Malloy family (326A), he married Mary Mooney (340), one of the better mid-wives, who died of an accident — she fell from the ladder while washing her windows — at the age of 108. H.G.

(327) Tom McCauley's. Dan's son, who continued the farm.

(328) Bassett's. Sheldon H. Bassett had a house here before 1900. His farm was straight in back of (257).

(329) Roosevelt School. This school was built in 1934 with materials brought across the ice straight from Charlevoix in cars pulling sleds. It was built on the site of the "Little Red School house," used from the 1880's as a school but in the 1920's as a dancehall before it burned down. L.J. Malloy, May Tilley, Irene Boyle, Vaughn Ogden, Pat Moran, Salty Gallagher, and Sister Leone taught at the first school, and Giles McCann at Roosevelt. Now a home.

(330) "Red Dan" Greene's. "Red Dan" was from Aran Mor. His daughter Elizabeth married Andy Mary Ellen.

(330A) Mike O'Donnell's. Anthony's cousin; he married Maggie Gibson and taught in the Sand Bay School (368). H.G.

(331) Anthony O'Donnell's. He came from Aran Mor by way of Canada with his wife Sophia (O'Donnell), the sister of Johnny the Rat (Black John Bonner's partner) and Paid een Og's (227) wife.

(332) Pat McDonough's. Pat, Vesty's son, started fishing at the age of eight and later learned smithing from Thomas Norton. He married Ellen O'Donnell, the daughter of Anthony (331), one of the first three Irish children born on the island. When the "Little Red Schoolhouse" burned (146) school was taught in a room of this house.

(333) Mike Burke's. His sons were Hugh, Phil, Tom (whose wife went blind), Mike, Ed, and Joe (who married a Maloney and bought the Gibson House in 1905). Coming from Connaught, they lived in a Mormon house here in 1859.

(334) "White Dan" Greene's. "White Dan," Johnny Green's father (270) — Johnny Green was the first Greene

to drop the 'e' — was the first of three generations of Greenes to use this farm.

(335) Paddy Ruah's. "Paddy Ruah" Gallagher was unrelated to all the other Gallaghers. He married a daughter of Thomas H. Boyle (274A); his son Hugh was known as the "Red Devil."

(336) Anthony — "Salty" — O'Donnell's — The Stone House. Called "Salty" because he made eleven trips across the ocean, he was no relation to the other Anthony O'Donnell (331), but his wife had previously been married, in Ireland, to still another Anthony O'Donnell. Later Jim Hamrock, who married Dan Boyle's daughter, built the stone house in front.

(337) Mike Gallagher's. "Mike Mahal Ruah" was a cousin of "Big Gallagher." He lost three daughters on the "Vernon."

(338) "Tight's" Hill — Dan Gallagher's. Dan Gallagher was called Tight, supposedly because when he came from New York in 1871 with his father he was wearing the fashionable tight pants. His father, James Peter, "Big Gallagher," a schoolteacher from Tyrone, was also the father of Bowery.

Big Gallagher and his son Tight's home (338).

(339) Manus Gallagher's. Another of Manus's homes. H.G.

(340) Owen Mooney's. Owen, from Aran Mor, was John Mooney's father, Barney Mooney's brother, and Mary Gallagher's husband. These Mooneys are unrelated to the other Mooneys (293). H.G.

(341) Manus Gallagher's. Manus was Cornelius's (295) cousin. H.G.

(342) J.C. "Salty" Gallagher's. Cornelius's (295) son was given this name because on a trip back to Ireland he learned to imitate the walk and talk of sailors.

(343) Neddy and "Big Biddy" Boyle's. The parents of "Red Huey," who came from Ireland through Canada. H.G.

(343A) Sunnyside School. From about 1875 to 1900 a log house here was used as a school. Margaret Gibson, Maggie Gordon, Mike O'Donnell, Elizabeth and Sally Dunlevy, James and Mel Gallagher, Frank McCauley, and Pat Boyle were the teachers. In 1900 Joseph Schmidt and Henry "Dutch" Hardwick built Sunnyside. Nuns assisted in the teaching, at first one for a full day and another for a half day. After 1914 Sunnyside students could go on to high school in St. James. In 1938 the eighth grade was transferred to the McKinley School. In 1945 the school closed, and in 1963 the building was made into a private home.

Map IV

(326A) Malloy's. Dan Malloy and his brother Jack had the beach property. Dan married Katcheline Mor; Anthony (110) was their son. H.G.

(344) Redding's Cabin. Mrs. Redding came to stay at the ill-fated Wildwood Inn. After living at "Little Marsh Hill" she moved to this cabin and lived alone in it until after WWII. The trail leading to it is called "Redding's Trail."

(345) Angelline's Bluff. Angelline Wabaninkee was a leader of the Peshawbes' Town Indians. Later she married Peter Taylor (65).

(346) Camp #2. Another B.I.L.Co. camp, on the main railroad line.

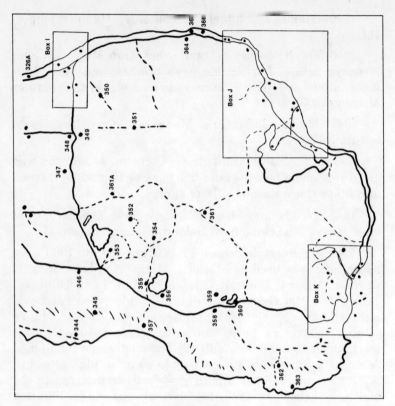

MAP IV

(347) Scopp (see 151 & 175) homesteaded this land, clearing two fields and starting a house.

(348) "Shamy" O'Donnell's. He claimed to be an heir to the O'Donnell Castle, in Ireland. H.G.

(349) Oil Well #1. The first of three 4000 foot deep "dry" wells sunk by the McClure Oil Company in 1961.

(350) From a mile south of (349) a wooden track — the Mill Road — on which oxen pulled carts of logs, angled northeast to Sweet's Mill (367).

(351) An unfinished extension was cut for use in reaching Lake Geneserath in the winter. There's an island legend that the Mormons had a glass mill on this trail.

(352) The Federal Lake Survey Tower. This is still barely erect.

(353) The State Fire Tower. Built in 1935, dismantled in 1964.

(354) Camp #3. Another B.I.L.Co. camp, reached by a railroad spur from "The Y," the corner of the south end of the Fox Lake Road and the West Side Road.

(355) Neil Greene lived in a large two-story house on the north bank, later used by Amesbury, the cook for camp #3. (H.G.)

(356) Tony Wojan's mill. In the early 1940's Tony Wojan ran his second mill here. The frame house a mile east of Tony's probably belonged to a homesteader.

(357) Green's Bluff. (see 203)

(358) Carpenter's Mill, Antrim Iron Works. From 1920-45 this was a base for the Antrim Iron Works, run by George Woods. They cut timber that was shipped to Mancelona, where it was made into charcoal. After the site had been unused for awhile the Carey Handle Company set up a mill to make rough parts for croquet sets, using some of the Antrim buildings.

(359) Oil Well #2. (see 349)

(360) Camp #5. This lumber camp and its jamway to the banking grounds a quarter mile southeast marked the southern end of the B.I.L.Co.'s narrow gauge railroad. The men's quarters were on the west side of the road, and the barns on the east.

(361) Doty's Camp. Doty was an independent logger who sold logs to the B.I.L.Co. His trail runs from Fox Lake Road to Miller's Marsh.

(361A) "Middie" Perron's. Amedie's barn was built over the road. He was the local berry-picking champion.

(362) Dan Boyle's. Dan Boyle and his son Dan had a fishing camp here (and another at Cheyenne Point). The young Dan bought the land in 1886. The bay was a fine harbor then but two big ice shoves ruined it, and the dock washed out before the turn of the century. In the 1930's Nels LaFreniere had a lumber camp here, using the Dan Boyle buildings. A three-master, the "Redford," came here once to take all the logs cut during that year; the "Rambler" took the logs out to her. One pile cut four inches too short was left on the beach, and can still be seen.

(363) The Powder House. Two men were poking at a keg of unfamiliar powder they came upon here; Dan Martin set it off, and died in the explosion. For a long time traces of the blast could be seen on the rocks.

(364) Gracie Martin's Hill. Gracie was Ed Martin's wife; she lived here long after Ed drowned in front of the house.

(365) Martin's Bluff. Mike, Dan, Ed, John, & Jim Martin lived close together here from 1857. John had two houses, one on the beach for summer fishing and another in the cedars for winter tie cutting. The first boat in the spring would buy his ties. Mike and Ed drowned, and Dan was blown up (363).

(366) Shoemaker's. The second home of John "Shoemaker" Gallagher, who fished pound nets here and had a small mill more than a mile above his house.

Box I, Hannigan's Corner

(367) Sweet's Mill. George Sweet bought a lot of land in the mid 1870's for Boardman and Sweet (Boardman had built the first mill in Traverse City), and had a mill here. H.G.

(368) A schoolhouse, used before the turn of the century. James "Ned" McCauley, a soldier in the Irish Civil War, taught here. H.G.

(369) Ned McCauley's. Ned was unrelated to all the other McCauleys. His wife was killed in 1891 by a runaway horse. Their son James taught school (368) and showed the students how to catch suckers with their hands in the Jordan River behind the schoolhouse. H.G.

(369A) Shoemaker's first house. He had a store here called "The Blue House." His father drowned, falling through the ice attempting to walk to the mainland. Shoemaker is said to have been a Supervisor of Galilee Township. He was also Hannigan's (224) half brother. H.G.

(370) "Big Dominick" Gallagher's. He was a brother of Big Owen and Big Neil. H.G.

(371) "Red Mike" Boyle's. Born in Pennsylvania, he was

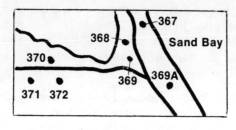

Box I

208

a half-brother to Hughie Boyle, and also to "Black Pete" McCauley. H.G.

(372) Hughie T. Boyle's. The father of Tom "Hannigan" and Peter Boyle, he married Hannah Beag who was Shoemaker's (369A) mother by a previous marriage. H.G.

Box J, Cable's Bay

(373) St. Ignatius Church. A small Catholic Chapel built by Father Murray around 1860 on land donated by the Martins and the Sullivans. After blood was shed in the building it was used as a school; Nora Gallagher (for 3 years), Nellie Gallagher (2), Mamie Maloney (4), and Rose Maloney (3), taught here following Margaret, Sarah, and Minnie Kilty (1 year each). It was discontinued in 1900 and razed by the Cole brothers in 1915.

(374) Sullivan's. John Sullivan fished and lived here. He was possibly related to the Pat Sullivan who fished here before the Mormons and was accused by Strang of having stolen his cow. H.G.

(375) Walter Smith's. He was a son of Smith the cooper. H.G.

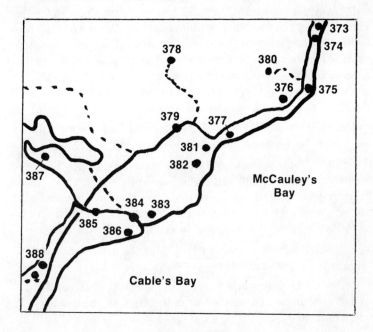

Box J

(376) Kilty's Hill. After Pat Kilty (211). People liked to boast that they'd been able to make Kilty's Hill in high in their cars; the first cars had to be turned around and backed up. (H.G., and hill almost gone.)

(377) Citizen's Realty Plat (from north of Kilty's Hill). This company sold lots, sight unseen and unseeable because of their vague descriptions, from 1910-1915.

(378) Bryan Gallagher's. Bryan was a brother of "Danny Don Mor" (282). Later Pfieffer acquired the land and planted the island's largest orchard, which was tended by Dave Wilson. Art and Gillis Larsen took the first crop to Petoskey in the 1920's. After operating very successfully for a number of years they were induced by an M.S.C. man to try a new spray, but it ruined the trees. Abandoned before W.W.II.

(379) Trombley's Hill. Trombley fished and did business with John Corlett, his father-in-law, on Cable's dock in 1857. He bought land here but sold it to James Cable in 1863. H.G.

(380) Burke's Orchard. This orchard was planted by a Burk, not related to any of the island Burke's. After he'd given up on it, though, Edward "Bub" Burke, who drove a taxi and carried the mail, bought it.

(381) Cole's Mill. About 1912 Garrett Cole bought the mill Sweitzer had built on land bought from Pete McCauley. There was a hardwood-floor dancehall on the dock. He threw Cole's Jubilee and so many people came by boat and buggy from all over the island that he followed the Jubilee with a succession of similar parties. He named the store and P.O. after his boat, the "Nomad." The mill ran until about 1926; the P.O. was open from 1917 to 1933.

(382) Pete McCauley's. Pete married Pat Malloy's daughter Ellen. Later the Nackermans lived here. H.G.

(383) Cable's Dock and Store. Steamers stopped here for cordwood in the 1840's and 50's. Two thirds of the men living here in 1850 were fishermen, and they caught enough fish to provide work for ten coopers. The Mormons renamed the Cable's Bay settlement "Galilee" and expanded it.

(384) John Johnson's. His first house straddled the stream here (see 217). H.G.

(385) The Mormons named this creek the "River Jordan" in 1852, a name that lasted past the turn of the century. It's now referred to as "Cable's Creek."

(386) Wildwood Inn. A luxurious hotel was started here

210

in 1916, designed by the architect of the Grand Hotel. By 1919 financial problems brought the construction to a halt, and that fall, after the summer guests had gone, the unpaid laborers began taking down and taking home the boards they'd put up. The end was marked by the death of a neighbor girl who set off a rigged gun. The absent watchman was tried for murder, and convicted, but Father Jewell had his sentence commuted.

(387) Hemlock Point.

(388) A Mormon Canal. A canal was started at Lake Geneserath intended to reach Lake Michigan, which was 26 feet lower, in order to place the drop at one spot to provide water power for Galilee. Work stopped with a third done.

Box K, Iron Ore Bay

(389) Keller's Hill. Keller ran a lumber camp on the hill for the Michigan Maple Block Company.

(390) Miller's Hill. After "Tip" Miller (see 232).

(391) Camp #4. Established by the Beaver Island Lumber Co. as their softwood camp, but not linked by their railroad. Most of the wood from this camp was taken down to Iron Ore Bay, and boomed to the mill in town by the tug, the "Ryan." Sometimes the booms were so big they weren't sure they'd be able to get them in the harbor.

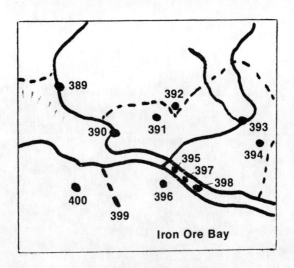

Box K

(392) Oil Well #3. (see 349).

(393) Iron Ore Creek. Flows into Iron Ore Bay (206).

(394) Leatherneck's. Leatherneck, who farmed here, was John Gallagher, a cousin of Bryan Don Mor (377). H.G.

(395) Mrs. Lasley's. She was an Indian midwife who made herbal medicines. Later Nicksau Plaunt, who was "ready to fiddle all night for a piece of salt pork" lived here, and the hill was called "Nicksau's Hill." H.G.

(396) Milwaukee Belle. She ran aground in 1886. Captain Roddy (296) bought her but was frustrated in his attempts to free her. First he hired a tug to tow her off, but the cable was too short. Several wagonloads of men came down and began lashing saplings together to extend the reach of the cable. They finished late at night and went to sleep, but when they awoke the season's first snow had fallen and the tug was gone. The winter's ice pushed her further up the beach. Every effort made for the next two years failed, and finally she caught on fire, or was set on fire. Roddy died of pneumonia he caught while trying to free her. Nothing remains but the keel in the sand near the beach close to the signal station.

(397) Built in 1851, the Beaver Head Light — and signal station (398) — is the third oldest light in the Great Lakes. The government had purchased land for the light, but it was built in the wrong place; it took the government 95 years to acquire the land under its buildings. The light ceased opera-

Beaver Head Signal Station and boat house (398).

212

tion recently when the Coast Guard decided that a small light on a new metal tower would be better. The French prism light (which was supposedly the second oldest in the U.S.) was discarded and then lost by the C.G.

(399) Michigan Maple Block Company Dock. Built in 1928 by John W. Green; railroad cars were pulled onto it, by horses on a tramway that ran to the base of Miller's Hill. Their steamer, the "Stuart," took their logs to Petoskey.

(400) "Betsy Smith." The wreck of this ship can still be seen in the water, loaded down with its cargo or ballast of iron ore.

The McKinley School (177).

1976